Artificial Intelligence, Co-Creation and Creativity

Artificial intelligence (AI) has deeply impacted our understanding of creativity and the human ability to generate creative outputs. New applications for creative tasks are rapidly evolving, and new tools are constantly being developed with much greater optimal capabilities. Importantly, the success of implementing such tools for creative tasks is still heavily dependent on human supervision and input. Therefore, it is vital to understand and critically reflect on the nature of co-creative processes between humans and AI. This book addresses such issues and provides insights into how humans can augment their capabilities for generating creative and innovative outputs by successfully co-creating with AI.

The book is intentionally divided into three main parts to allow for a comprehensive and holistic perspective on human and AI co-creation for creative tasks. The sections are divided as follows: Part 1: "Principles of AI and Creativity", Part 2: "Critical Issues on Artificial Co-Creation", and Part 3: "Industry-Specific Discussions".

Consequently, the book provides a holistic insight on the topic, covering various issues and perspectives and enabling an accessible read to a broad audience. For example, chapters cover examples across different industry sectors, including music, arts, science, and management. Furthermore, the book covers critical questions involving copyrights, ethical concerns, relationship with algorithms, and context-based issues. Only by critically reflecting on the intrinsic issues of AI and learning how to work with it effectively for creative purposes will we be able to benefit from its full potential to augment human creative abilities in an appropriate manner.

This novel, edited collection is an essential read for scholars working on the intersection of AI, creativity, arts, and management.

Francisco Tigre Moura is Professor of Marketing at the IU International University of Applied Sciences, Germany.

RIOT!

Routledge Studies in Innovation, Organizations and Technology

Digital Business Strategy
Content, Context and Cases
Anders Pehrsson

Cyber Security and Business Intelligence
Innovations and Machine Learning for Cyber Risk Management
Edited by Mohammad Zoynul Abedin and Petr Hajek

Internet of Things in the Food Industry
Challenges and Opportunities for the Internet of Food Things
Edited by Anna Rogala, Renata Nestorowicz and Ewa Jerzyk

Corporate Entrepreneurship and Innovation in Tourism and Hospitality
Global Post Covid-19 Recovery Strategies
Edited by Teresa Aguiar-Quintana, Jonathon Day and Francisca Rosa Álamo Vera

Grassroots Innovation
Discourse, Policy and Practice in the Global South
Hemant Kumar and Gautam Sharma

Business Models and Digital Technology Platforms
Implementation and Complexities for Digital Business
Krzysztof Bartczak

Artificial Intelligence, Co-Creation and Creativity
The New Frontier for Innovation
Edited by Francisco Tigre Moura

For more information about this series, please visit: www.routledge.com/Routledge-Studies-in-Innovation-Organizations-and-Technology/book-series/RIOT

Artificial Intelligence, Co-Creation and Creativity

The New Frontier for Innovation

Edited by Francisco Tigre Moura

Routledge
Taylor & Francis Group
LONDON AND NEW YORK

First published 2025
by Routledge
4 Park Square, Milton Park, Abingdon, Oxon OX14 4RN

and by Routledge
605 Third Avenue, New York, NY 10158

Routledge is an imprint of the Taylor & Francis Group, an informa business

British Library Cataloguing-in-Publication Data
A catalogue record for this book is available from the British Library

ISBN: 978-1-032-59262-6 (hbk)
ISBN: 978-1-032-59263-3 (pbk)
ISBN: 978-1-003-45390-1 (ebk)

DOI: 10.4324/9781003453901

Typeset in Sabon
by Apex CoVantage, LLC

"Dedicated to my son Leonardo, and to the loving memories of my father Romero and my mother Teresa".

Contents

Figures

Tables

Contributors

Pasha Abkenar is a recent graduate in international marketing management from the IU University of Applied Sciences, Germany. Currently, he is involved in the DeutschAkademie, Hamburg, actively working in the pursuit of learning, exploring, and experimenting with generative AI.

Claire Anscomb is a practicing artist and a lecturer in fine art at De Montfort University. Prior to this, she was the 2021–2022 British Society of Aesthetics postdoctoral fellow in the Philosophy Department at the University of Liverpool and was awarded a PhD in history and philosophy of art from the University of Kent in 2019. She has been the recipient of awards for her artistic and philosophical work, including the 2014 Signature Art Prize in Drawing and Printmaking and the American Society for Aesthetics 2021 John Fisher Memorial Prize, and has published on topics including AI art, creativity in artistic and scientific practices, and photographic phenomenology.

Jonathan Attard is a student currently pursuing a master's degree in AI with a specialization in computer vision and media. He has experience working with AI models and aims to contribute to the ongoing development of the field. Outside of his technical pursuits, Jonathan enjoys exploring the creative realm with a background in music, having engaged in music theory and casual piano play for leisure over the years.

Mirjam Augstein is a professor of personalized and collaborative systems at the University of Applied Sciences Upper Austria in Hagenberg. Her main research interests are rooted in the broader field of HCI, focusing on computer-supported cooperative work and hybrid collaboration as well as personalized interaction.

Trevor Borg is a multi-disciplinary artist, curator, and academic and the head of the Department of Digital Arts at the University of Malta. His practice traverses place and space, delves into ambiguous territory, and oscillates between the poetic and the political. His work spans a wide range of media, from drawing, painting, and sculpture to film and installations. His work has been shown in a number of prestigious venues worldwide, including the Venice Biennale 2019, the Beijing International Art Biennale 2022, the London Design Biennale 2023, and in many other exhibitions internationally.

Karin Breckner is a researcher at the University of Applied Sciences Upper Austria in Hagenberg and a PhD student of computer science at the Johannes Kepler University in Linz. Her research focus is on human–AI collaboration from a computer-supported cooperative work perspective.

Oreste Leone Campagner is a theatre practitioner and independent scholar. He is a proud alumnus of the Ghislieri College in Pavia, Italy, and of the Ecole Normale Supérieure in Paris, France, where he obtained a MSc in theoretical chemistry. He holds a bachelor's degree in acting from the ERT National Theatre in Modena and is currently studying philosophy at the University of Bologna, Italy. His interests lie at the interface between sciences, philosophy, and the arts.

Nicholas Davis graduated from Case Western Reserve University with a BA in cognitive science. He received his PhD in human-centred computing from the Georgia Institute of Technology, where he specialized in cognitive science and computational creativity. His thesis work investigated methods of quantifying co-creative AI experiences. He spent four years as an assistant professor at the University of North Carolina at Charlotte. He is currently working as a research consultant at Co-Creative AI Consulting.

Mariya Dzhimova is a research associate at the Institute for Culture Management and Media at the University of Music and Performing Arts Munich and head of the digitalization lab "Artificial Intelligence in Culture and Arts" (AICA).

Stefan Ellenberg specializes in intellectual property, copyright, and media law. He currently serves as the head of business affairs at Gutsch & Schlegel Rechtsanwälte, providing legal guidance to both national and international clients, primarily within the domains of music, film/TV, and merchandising/licensing. In addition to his role, Dr Ellenberg is an active contributor to WEB3LEX, a web3 information resource. He consistently publishes articles and conducts webinars addressing various aspects of intellectual property law.

Thorsten Fröhlich has had a distinguished career both in academia and business. Academically, he graduated with a degree in chemistry from the University of Cologne, where he earned his doctorate. He has held various academic positions, including guest lectureships and professorships, with his current role being a professor of IT management and Big Data at the IU University of Applied Sciences. In the business world, he has been a founder and managing partner in several companies, notably in scientific publishing, software development, and marketing informatics. Currently, he serves as the managing partner of Life Science Network Beteiligungsgesellschaft GmbH & Co KG in Berlin and the chairman of the Supervisory Board of LUMITOS AG.

Cansu Hattula is a professor at the IU International University (Hannover, Germany). She completed her studies in business administration at the University of Hannover (Germany) and her PhD at the University of St Gallen (Switzerland). Her research interests include change management and marketing strategy. She has worked in various organizations and start-ups across Germany, Switzerland, and the UK.

Markus C. Hemmer began his scientific career as a chemical engineer and worked in analytical geochemistry at the universities of Mainz and Hanover. He studied chemistry in Hanover and Erlangen and completed his doctorate in computational chemistry on mathematical models for the digital representation of molecules. After years in research, he moved to the software industry as a product innovation manager for scientific information systems. He is currently a professor and the programme director for IT management at the IU International University.

Clare Hindley is a professor of language and communication at the IU International University of Applied Sciences, Germany. She holds a PhD in sociolinguistics and has had a diverse international career in business and education. Her current work focuses on business and management, with particular interest in the fields of human perception and education. Her research is cross-disciplinary, with recent publications in the areas of creativity, culture, communication, and sustainability.

Frederik Hirschmann is a researcher and lecturer at the University of Applied Sciences Upper Austria in Hagenberg and a PhD candidate of Computer Science at the Johannes Kepler University Linz on the topic of hybrid collaboration. His main research interests are rooted in the broader field of HCI, focusing on computer-supported cooperative work and hybrid collaboration.

André Hollstein is the director of digital transformation at EBS University, a former professor of information systems at the IU International University and CIO of business stream mobility at TÜV Rheinland. His work and research focus on digital transformation, software engineering, and digital education. He regularly publishes episodes on current digitalization topics on DigitalXRoad.com.

Kristina Khutsishvili is a postdoctoral researcher at the Informatics Institute, University of Amsterdam, and the 'Ethical and Inclusive Engagement in Practice' work package leader of the CommuniCity Horizon Europe project, a collaborative project with the cities of Amsterdam, Helsinki, and

Porto aimed to bring together the world of technologies and vulnerable and marginalized communities by means of co-creation.

Nicole Lontzek was born in Augsburg and studied at Munich University of Applied Sciences and the I.E. Business School. Her professional career has taken her to international tech companies, such as Google in Dublin, as well as to deep tech AI start-ups. The marketing expert specializes in the marketing of complex B2B software products. She is currently the marketing director of the Central European business at Dataiku.

Caterina Moruzzi is a chancellor's fellow in design informatics, School of Design at the University of Edinburgh, working at the intersection between the philosophy of art and history, the philosophy of human and artificial creativity, and the philosophy of AI. In her ongoing projects, she investigates modes of shared agency and creativity between humans, data, and technology and the disruptive effects that emerging technological innovations can have on creative labour.

Francisco Tigre Moura holds a PhD in marketing from the University of Otago (New Zealand) and has worked since 2013 as a professor of marketing at the IU International University of Applied Sciences, Germany. He has multiple publications in scientific journals, conferences, and specialized media and is the founder of LiveInnovation.org, a research and education website which, among other topics, discusses the use of AI in the music sector. He also leads the monitor project "Live FM: Fan Monitor", which investigates behavioural trends of music fans in Germany, including perceptions towards AI in music. He also often attends industry and academic events as a presenter and moderator to discuss artificial creativity.

Tobias Naujoks has been a professor of business administration, in particular marketing, at the IU International University of Applied Sciences since 2021. He has many years of management experience in the areas of marketing, sales, and analytics and has worked for leading global companies in the consumer goods industry and the retail/technology sector. Previously, he studied business administration in Bonn, Paris, and Leipzig and completed his doctorate at the Chair of Marketing Management and Sustainability at HHL Leipzig Graduate School of Management.

Thomas Neumayr is a researcher and lecturer at the University of Applied Sciences Upper Austria in Hagenberg. He earned a PhD in business informatics at the Johannes Kepler University Linz on the topic of hybrid forms of collaboration. He teaches courses on human–computer interaction, adaptivity, and personalization.

Ali Nikrang is a multidisciplinary AI researcher and artist in the fields of music and artificial intelligence (AI). He is a professor of AI and musical creation at the University for Music and Theatre in Munich and a researcher at the Ars Electronica Futurelab in Linz. His main research interest lies in the creative applications of AI in the field of classical music.

Elena Said recently graduated with a master's degree in digital arts from the University of Malta. Drawing on her background in art history, she has enriched her understanding of art, which now plays a pivotal role in her exploration of digital art. Her dissertation, focusing on the integration of AI in art creation, resulted in an impactful installation showcased at Spazju Kreattiv, Valletta. Passionate about research, she continues to leverage her knowledge to push the boundaries of artistic expression in the digital realm.

Benjamin Schulte currently serves as a professor of marketing management at the IU International University Berlin Campus. Prior to his professorship, he worked in consulting for B2B sales and project management. His research interest is focused on customer-oriented digital systems in retail and communication, as well as their strategic backgrounds and implications.

Dylan Seychell, an AI and computer vision academic at the University of Malta, has focused his scholarly attention on AI safety and usability by publishing several peer-reviewed publications. He is a technical expert certified by the Malta Digital Innovation Authority and was a Maltese government AI task force member who drafted the National AI Strategy in 2019. He founded Colour My Travel and Beyond Museums and is trusted in the industry with several high-profile projects that apply AI and high-tech to tourism and cultural heritage. These include the Caravaggio Wing, the Augmented Reality Experience at the Sacra Infermeria, and the National AI Tourism Platform.

Elif Surer is currently an associate professor in the Graduate School of Informatics, Department of Modeling and Simulation, Middle East Technical University (METU), Ankara, Turkey. Her research interests include serious games, virtual/augmented reality, human and canine movement analysis, machine learning, and computer vision.

Acknowledgements

First, I would like to express my deepest gratitude to all the authors who have contributed to this book. It has been a true honour collaborating with you on this project, learning from your expertise and being truly inspired by your ideas and insights on the fascinating world of "Artificial Intelligence, Co-Creation and Creativity".

I would also like to thank Alex Atkinson, Manjusha Mishra, and the entire Routledge team for their trust and encouragement. It has been great working with you!

Finally, I would like to thank my colleagues for their support, other academics in the field for the inspiring work they have done (we are all standing on the shoulders of giants!), and industry innovators who constantly develop the latest technologies that shape the societal transformation we are witnessing, being impacted by and researching.

Introducing *Artificial Intelligence, Co-Creation and Creativity*

The New Frontier for Innovation

Francisco Tigre Moura

The emergence and proliferation of artificial intelligence (AI) tools for creative tasks is deeply impacting and rapidly transforming various creative sectors, allowing for a permanent state of innovation across different domains. Such fast-changing scenario challenges established assumptions and calls for a constant critical reflection among scholars and practitioners. This first edition of *Artificial Intelligence, Co-creation and Creativity: The New Frontier for Innovation* aims to explore the interlinks between humans, AI and creativity and the potential for innovative co-creative processes. It also aims to contribute to the current debate by providing a holistic insight on the topic, covering various issues and perspectives and enabling an accessible read to a broad audience.

Creativity in Times of AI

Creativity, once thought to be an exclusive human trait and essential part of a sense of self identity, must now be redefined. Traditionally, creativity has been viewed as a process of producing novel and valuable ideas or solutions (Boden, 2004), often associated with the genius of individuals. However, this paradigm often does not account for the autonomous and impersonal nature of algorithms and the increasing advancements in AI's capabilities which challenge this perspective even more.

The rational or abstract applications of AI (e.g. for artistic or scientific purposes), whether for composing music (Lopez-Rincon et al., 2018), writing novels (Ghajargar et al., 2022), or identifying original insights to enable a scientific discovery (Wang et al., 2023), have provoked scholars and society in general to question what truly constitutes creativity and to reflect what the role of humans in this innovative process is. Also, a thorough investigation of the perception towards artificially created or co-created outputs can be seen in current research (Ragot et al., 2020; Tigre Moura & Maw, 2021; Tigre Moura et al., 2023).

DOI: 10.4324/9781003453901-1

Moreover, humans have often adopted smart systems as a cognitive assistant tool, with the aim of benefiting from their capacities of scale for a productive self enhancement. But with the rapid developments of machine learning, neural networks, and deep learning techniques, AI has transcended from being just a passive tool to an active collaborator in the creative process and is often also perceived as an independent creative entity, possessed with its own intentionality. This has led to completely innovative co-creative processes, where the role of humans in creative tasks must be reconsidered.

Such drastic change results in the need for the development of a new paradigm for creativity and creative processes (Tigre Moura, 2023; Vinchon et al., 2023). Thus, it is time for a deep critical reflection, to be open for establishing and accepting new perspectives on creativity, and to adapt to the fast-moving reality of artificial creativity.

As a society, we must account for and embrace a new perspective on creativity where humans and algorithms foster innovation together, in various possible modes and levels of co-creation. This also includes accepting the creative outputs of non-creatives, changing our viewpoints on the role and relevance of creative expertise and skill, and accepting new creative skill forms, such as mastery of smart system tools. It is indeed a new era for creativity.

However, every groundbreaking innovation also creates challenges, which are resultant from a disruptive scenario. Despite the inspiring prospects that come along from an ever-evolving human–AI co-creation relationship, there are serious issues which must be addressed. Adaptation to such a revolutionary shift calls for an in-depth understanding and thoughtful considerations. For example, artificial creativity poses a profound impact on human sense of creative self, the value of creative products, and new and unthought challenges in the legal arena. As human–AI co-creation with smart systems becomes even more widespread, questions and insights involving authorship, originality, and value of AI-generated work become even more important (Yanisky-Ravid & Velez-Hernandez, 2017; Tigre Moura et al., 2023).

Thus, *Artificial Intelligence, Co-creation and Creativity: The New Frontier for Innovation* aims to contribute to the current discussions in the field, by enabling academics and practitioners to analyse the complexities of these issues. The chapters reveal valuable insights to potential streams of thought in the realms of AI, creativity, and co-creation, and all chapters conclude by providing clear directions for future research in the field. Thus, it both informs and inspires readers to navigate through the new frontiers of innovation, how AI is augmenting human creative capabilities, reshaping societal structures and industry dynamics and reimagining the creative field.

Structure of the Book

The book is intentionally divided into three main parts, to allow for a comprehensive and holistic perspective on human and AI co-creation for creative tasks. The sections are divided as follows:

- Part 1: Principles of AI and Creativity
  This initial section provides an overview of AI and its core concepts, allowing readers an in-depth understanding of topics and discussions throughout the chapters. Moreover, the section also discusses human and AI co-creation and provides a critical reflection on creativity in the times of AI, the need to reconsider existing concepts, frameworks, and paradigms, as well as its potential to enhance human creative capabilities.
- Part 2: Critical Issues on Artificial Co-Creation
  The second part is devoted to addressing specific issues which arise as a consequence of human and AI collaboration for creative tasks. They include, for example, copyright, the measurement, and definition of quality of AI co-created outputs, a critical reflection on human creative augmentation, and even a rethink of what constitutes being human. In sum, the section reflects on ways to overcome legal and ethical barriers, conceptual concerns in the relationship between humans and algorithms in the creative realm.
- Part 3: Industry-Specific Discussions
  This final section aims to trigger reflection on how the application of human and AI co-creation must be considered in light of the specificities of the industry sector the collaboration is made in. Chapters delve into sectors such as arts, science and research, marketing, and gaming. These different perspectives will help readers understand the need to adapt to context-related issues while co-creating with smart systems, thus hoping to help mature the adoption of artificial co-creation.

Final Remarks

Despite the challenges of the creative disruption led by the autonomous and co-creative process with smart systems, human creativity will not be diminished. Instead, it must be viewed as in the process of immense augmentation, which will lead to an abundance of creative outputs. Artificial creativity will allow humans to reimagine and reinvent our creative capabilities, projecting creatives and non-creatives into a future of unimaginable innovation.

Developing these innovations and directing them towards creating meaningful impact should be the focus of most of our creative actions. To reflect on the intersection between *artificial intelligence, co-creation, and creativity* is to embark on a journey that uncovers not just the innovative potential of these synergies but also the extremely relevant issues they inevitably raise which will impact various industries and society.

Finally, it is my hope that this book, and the various perspectives raised along the different chapters, will provoke thought and discussions, stimulate research at all academic levels, and inspire action, helping humans to better understand and shape the future in the landscape of creativity.

References

Boden, M. A. (2004). *The creative mind: Myths and mechanisms*. Psychology Press.

Ghajargar, M., Bardzell, J., & Lagerkvist, L. (2022, November). A redhead walks into a bar: Experiences of writing fiction with artificial intelligence. In *Proceedings of the 25th international academic MindTrek conference* (pp. 230–241). https://www.researchgate.net/publication/365440278_A_Redhead_Walks_into_a_Bar_Experiences_of_Writing_Fiction_with_Artificial_Intelligence

Lopez-Rincon, O., Starostenko, O., & Ayala-San Martín, G. (2018, February). Algorithmic music composition based on artificial intelligence: A survey. In *2018 international conference on electronics, communications and computers (CONIELECOMP)* (pp. 187–193). IEEE.

Ragot, M., Martin, N., & Cojean, S. (2020, April). Ai-generated vs. human artworks. A perception bias towards artificial intelligence? In *Extended abstracts of the 2020 CHI conference on human factors in computing systems* (pp. 1–10). CHI.

Tigre Moura, F. (2023). Artificial intelligence, creativity, and intentionality: The need for a paradigm shift. *Journal of Creative Behavior*, *57*, 336–338. https://doi.org/10.1002/jocb.585

Tigre Moura, F., Castrucci, C., & Hindley, C. (2023). Artificial intelligence creates art? An experimental investigation of value and creativity perceptions. *The Journal of Creative Behavior*, *57*, 534–549. https://doi.org/10.1002/jocb.600

Tigre Moura, F., & Maw, C. (2021). Artificial intelligence became Beethoven: How do listeners and music professionals perceive artificially composed music? *Journal of Consumer Marketing*, *38*(2), 137–146.

Vinchon, F., Lubart, T., Bartolotta, S., Gironnay, V., Botella, M., Bourgeois-Bougrine, S., Burkhardt, J.-M., Bonnardel, N., Corazza, G. E., Glăveanu, V., Hanchett Hanson, M., Ivcevic, Z., Karwowski, M., Kaufman, J. C., Okada, T., Reiter-Palmon, R., & Gaggioli, A. (2023). Artificial intelligence & creativity: A manifesto for collaboration. *The Journal of Creative Behavior*, *57*.

Wang, H., Fu, T., Du, Y., Gao, W., Huang, K., Liu, Z., Van Katwyk, P., Deac, A., Anandkumar, A., Bergen, K., Gomes, C. P., Ho, S., Kohli, P., Lasenby, J., Leskovec, J., Liu, T. Y., Manrai, A., Marks, D., Ramsundar, B., . . . Zitnik, M. (2023). Scientific discovery in the age of artificial intelligence. *Nature*, *620*(7972), 47–60.

Yanisky-Ravid, S., & Velez-Hernandez, L. A. (2017). Copyrightability of artworks produced by creative robots, driven by artificial intelligence systems and the concept of originality: The formality-objective model. *Minnesota Journal of Law, Science & Technology*, *19*.

Part 1

Principles of AI and Creativity

Chapter 1

AI Creativity in the Light of Autonomy

Ali Nikrang, Karin Breckner, Thomas Neumayr, Frederik Hirschmann, and Mirjam Augstein

Introduction

Art and creativity count as two very human abilities. Due to the ability to develop new ideas, share these ideas with others, and pass them on to the next generations, humans have been able to develop tools and technologies that give it capabilities far beyond its physical capabilities. Moreover, the ability to create visual and auditory or narrative impressions and to share these impressions with others through painting, storytelling, or musical performance owes much to the creative imagination of humans, which has been documented in cave paintings and archaeological findings (e.g. bone flute) at very early stages of human development.

Technology has always played a role in the development of the arts (e.g. through new musical instruments or better ways to make paints for paintings). In the previous 200 years, this process has accelerated, resulting in various technologies that have made new art forms possible. Photography, electronic music, and movies are just a few prominent examples.

With many older technologies, however, artists always had control over the outcome of the process. They themselves could accurately forecast how the equipment would react to their actions. They knew, for example, what effect a particular lens would have on a camera.

However, over the past years and with the upcoming of creative artificial intelligence (AI), the art community has been confronted with a new set of technologies that have provided completely new possibilities for artists. As a result, numerous projects have been developed in almost every field of art to demonstrate the creative potential of AI (see, e.g., Hieslmair, 2002; McKenzie, 2022; Unsupervised, 2022). Recent AI technologies acquire the capability to imagine and create completely new landscapes, combine different emotional and conceptual ideas by simply providing it with a text as a prompt. At the same time, it can be used to create stories and pieces of music that are capable of having an emotional impact on humans. It is expected that this development will have a significant role to play in many creative processes and art forms in the near future (Gownder et al., 2023).

DOI: 10.4324/9781003453901-3

Advances in the merging of AI, creativity, and art have been met with both excitement and discomfort, leading to a host of questions in artistic communities about the role of humans in AI creative processes. The question of how the role of humans in this creative process is defined is also an essential indicator to answer the question of ownership of works co-created with AI.

This chapter addresses the issue of human role in a co-creative process with AI. The focus will be on artistic creations. The chapter is structured as follows. First, it describes the crucial importance of the concept of co-creation for the use of creative systems in the field of arts. Then, the concepts of creativity, AI-based creativity, intention, autonomy, and collaboration will be discussed from different perspectives but in the context of co-creation of artistic works. In the second part of the chapter, a simple framework to determine the role of human in a creative co-creation space will be proposed. In the discussion section, different use cases involving different ways of interaction with the AI system will be discussed.

Human and Social Context of Arts

Arts and music can be considered as social phenomena. Many artworks appear in a certain cultural and social or political context. This context is an essential part of how the artwork is perceived by the society. Sometimes the context is obvious and explicitly described (e.g. by the title of the artwork or the lyrics of a piece of music such as Benjamin Britten's War Requiem or the lyrics of Beethoven's Ninth Symphony), sometimes it is not apparent at first glance. Historical artworks, for example, always have such a historical context from today's perspective, representing a certain time in a certain geographical cultural area, often combined with biographical details of the artist. Furthermore, art is a phenomenon that communicates on different levels. One of these levels regards the artists themselves. Popular artists or styles lead to the formation of communities in society. Tarr et al. (2014) show the impact of music in social bonding in an evolutionary perspective. From a socio-historical point of view, the sociologist Howard S. Becker defines *the artist* as the person who performs the core activity, without which the work would not be art and as the central figure in a network of cooperating people (Becker, 1982). He also underlines the role of the individual artist's reputation from different angles in the various fields of art that, beside other factors, shapes the way how artworks are perceived by the society. Given these socio-historical perspectives, it is safe to assume that human artists will always remain an essential component of a work of art, even in the age of AI, at least in most areas of art where the artistic aspect is paramount. Another reason for this assumption is that beside the socio-historical perspectives, also the creative act itself, the process of creation with an AI system requires an active role of human counterparts.

In the following section, various aspects of creativity are described with a focus on the actual performer of creation, *the artist*, followed by a section describing the act of creation from a technological perspective.

Human Creativity

There are numerous definitions of creativity. Margaret Boden, for example, defines creativity as "the ability to come up with ideas or artefacts that are new, surprising, and valuable" (Boden, 2004, 2007). In another definition, Csikszentmihalyi (1997) pays special attention to the person who performs the creative act and defines creativity as the interaction of a system of three components: *domain*, *field*, and individual *person*. The *domain* consists of rules and procedures; music and mathematics are such examples. The *field* further includes any person acting as a "gatekeeper" to the field (Csikszentmihalyi, 1997). This involves experts in the field who judge whether or not a new work should be recognized. In the age of social media, this can also be heavily influenced by recommendation systems and the inherent dynamics of social media. The third component is the *person* who creates a new idea (or artefact) using the symbols and rules defined in the domain. In the field of arts, it is the artist whose involvement makes the result to be accepted as art. The more famous the artist, the more attention the result will receive and the more likely it is to be recognized by the field. While Boden focuses mainly on the outcome of a creative act, Csikszentmihalyi's definition also includes actors from the field and society and, implicitly, the audience. In the field of arts, the person (or the artist) is him- or herself a context in which the result takes place. When the artwork is made accessible to the public, everything that is associated with the person, and known to the public about him or her, becomes part of the perception of the artwork. To give an example, regardless of whether one looks at a painting by Monet or listen to a Beethoven symphony or a song by the Beatles, a certain social and personal context of the artist and his or her time is part of the perception. Such a context might have been merely assumed (and not based on fact) or even be based on the audience's previous experience with similar artworks and artists. The field or experts in each case have a much more detailed and larger context about the person and build a social environment around the artwork. For both modern and historical artistic personalities, this can lead to a cult-like status among artists, where their influence and appeal goes beyond their work, often leading to passionate admiration, intense study of their lives, and a lasting impact on their field. (Heinich [2003] illustrates nicely this historical-social aspect, referring to the famous case of Vincent Van Gogh.)

AI Creativity

Modern deep learning systems are trained with large datasets, which facilitates their ability to fulfil a predetermined task (for an overview, see LeCun et al., 2015). The nature of the task varies depending on the generative AI system used. Autoregressive models, for example, can be used for sequence generation. These models aim to forecast the next element in a sequence based on previous elements. To fulfil this task, the model must learn an understanding of the underlying data to forecast the next element, given the context in which it will occur. A key advance for such models came with the introduction of Transformer technology by Vaswani et al. (2017). This concept is mainly used in language models (Radford et al., 2018; Zhao et al., 2023) and also often for symbolic music generation (e.g. Payne, 2019). In addition, its adaptability enables the generation of audio data, for which the model learns to generate the next audio sample token given the sample tokens before (Agostinelli et al., 2023; Dhariwal et al., 2020). In the visual data domain, diffusion models have proven to be effective tools. These models add iteratively Gaussian noise to the images and then iteratively denoise them, creating a detailed impression of the visually discernible structures during this process (Sohl-Dickstein et al., 2015). Generative Adversarial Networks (GANs) represent a distinct paradigm in generative modelling (see Goodfellow et al., 2014). GANs consist of a generator and a discriminator. The task of the generator is to produce images, while the task of the discriminator is to evaluate their authenticity against real images. This process requires a complex understanding of the real data distributions to distinguish between real and synthetic images. At the same time, the generator must learn the ability to produce increasingly convincing synthetic images so that it becomes more difficult for the discriminator to detect a fake image.

The result of the training is a model consisting of millions, or billions, of weights and biases which are numeric values that have been adjusted during the training. These values represent the statistical features of the data and are represented as multidimensional vectors in different layers of the model. These features reflect the whole dataset the model has been trained with. These multidimensional spaces are then used to generate data that are similar, though not identical to the data in the training set.

Autonomy

In human interaction with AI systems, *autonomy* describes the ability to not only make independent decisions, which is essential for creativity (Cetinic & She, 2022), but also to do so willingly (Calvo et al., 2020). Two main forms of autonomy are relevant to AI creativity: *executive autonomy* concerns the decision of *how* to reach a defined goal, whereas *goal autonomy* includes the definition of the goal itself (Castelfranchi, 1994). According to Jennings (2010), systems are required to act autonomously with respect to both goals and execution, yet not at random, to be considered creative. In human–AI co-creation (see below), the forms of autonomy may be shared between humans and AI to varying degrees, to shape the final creative output. Guidance may steer the collaboration to maintain the artistic vision and intended conveyance, despite the unpredictable nature of the creative AI's output. This ensures the artist's role and supports the complementary creation of artwork.

Collaboration

In humans, an activity or interaction is defined as collaboration when multiple persons are working together and several prerequisites are fulfilled, among them the mutual possibility to *communicate*, *contribute*, *coordinate*, and *cooperate* (Shah, 2010). From a human perspective, all these prerequisites can already be fulfilled to a certain extent by an AI counterpart, which would in turn lead to the conclusion that working together with an AI-driven system could theoretically be perceived as *collaboration* (e.g. an NPC companion in a role-playing computer game). However, several authors stress the fact that collaborators must *synergistically* work towards a shared goal (see Denning & Yaholkovsky, 2008; London, 1995; Shah, 2010). Most prominently, they hint at necessary intrinsic motivations with underlying goals, or intents, likely driven by the individual personal circumstances. It is still up to debate to which extent an AI in co-creation can be regarded as a "*true*", "*original*", or "*natural*" collaborator compared to the definitions that have been used for human-to-human collaboration. For example, as early as 2004, Klein et al. listed ten factors that have to be fulfilled for automation to be a "*team player*" in joint activities, including a certain commitment to goal alignment, mutual predictability, or the ability to observe status and intention. According to a more recent estimation by Abbass (2019), most of these factors are still not sufficiently considered by current AI systems.

From an artistic perspective, it might, however, be sufficient for the artist to merely *perceive* an AI as collaboration partner to co-create a work of art. The artist's intention is typically what gives meaning to the produced artwork (Jucker et al., 2014) and, at the same time, is difficult to express and explain to others. As of today, AI does not (yet) have the capability for personal intention. While sharing a mutual goal of producing a piece of art with an AI system may be possible, sharing the idea of what to convey with it is even more challenging yet crucial for the outcome to be perceived as art (Jucker et al., 2014).

Human–AI Co-Creation

Collaboration with a creative AI system can occur in a variety of ways. *Prompting* is a simple type of interaction where the user starts a sequence (typically language) and the system continues the sequence. Prompts simply initialize the context in which the model starts to interact. In further steps, the model's output will also serve as the context. For large language models (LLMs), prompts can also be used to cause the model to perform a task, for example, "*translate the following sentence into German*". An advanced form of prompting is recently developed LLM-based chatbots such as ChatGPT (2022). User prompts, while interacting with the model, continuously change the context of the conversation with the model, leading to an interactive conversation. Prompts are also used to conditionally generate and change the content of images. Users can use prompts that describe the context, style, and other features of a desired image, whereas the model comes with several suggestions from which the user can choose one. It is also possible to refine the generated images by adding more prompts. In the field of music composition with symbolic data, programs like Ricercar Nikrang (2020) or MuseNet Payne (2019) can use a musical prompt (a sequence of notes) to tell the model how to start a composition. In these cases, the model offers different possibilities of how the piece could continue in the next section (which can last between 10 and 30 seconds, depending on the tempo). The user can select one of the musical suggestions and let the model continue, whereupon the model suggests the next five suggestions for the next section, and so on.

Discussion

Up to this point, some of the special conditions that prevail in a creative and artistic environment shared with AI have been explained. Figure 1.1 provides an overview of several aspects discussed so far and their relation to the interaction with creative and non-creative systems. In the following, the role of the human creator in such an environment will be examined on the basis of three aspects: *actors*, *goals*, and the process of *co-creation*.

Actors

One aspect that most definitions of creativity directly or indirectly share is the role of an actor (for an overview, see Wingstroem et al., 2022). Ideas are

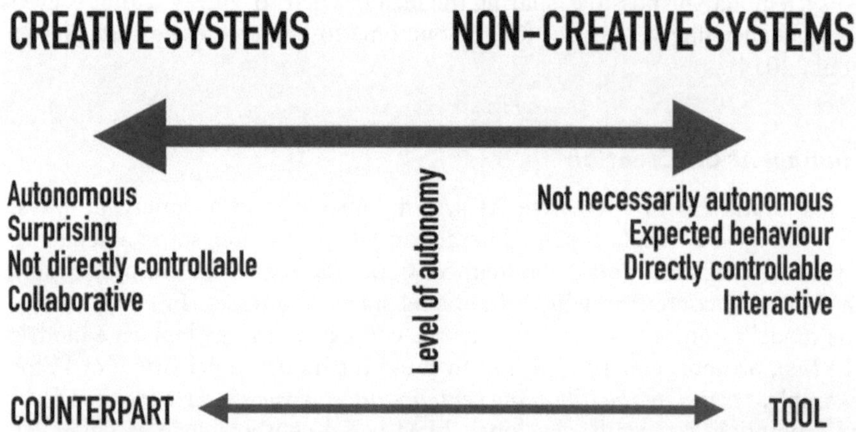

Figure 1.1 The figure shows the interaction between creative and non-creative systems as tools and counterparts.

created out of intent in an active process. For example, in Boden's (2004) definition of creativity, the "*person*" is not explicitly mentioned but can be implicitly understood as the agent who takes an active role in the process of creation to actively develop new ideas and artefacts. On the other hand, Boden's definition, implies creative systems to be autonomous systems as they must be capable of making decisions autonomously to produce data that are unexpectable and surprising (in addition to being new and valuable) – surprising results are not predictable. They require a certain degree of autonomy within the system. This contrasts with regular tools and systems that are used in the process of creating artistic results. For example, usage of traditional (non-creative) tools in applications, such as Photoshop or Logic Pro, comes with a certain expectation. When users apply a black-and-white filter in Photoshop or a low-pass filter in Logic Pro, they have a clear understanding and expectation of the result. However, when it comes to creative systems, this kind of expectation is misplaced: the more the result of a creative system can be expected, the less creative the system is considered. Following the aforementioned explanations, it is reasonable to divide the technical systems involved in the production of artworks into two categories: *non-creative systems* (that can be considered as *instruments*) and *creative systems* (that act as *counterparts*).

Instruments are non-creative systems that produce results that are meant to be expected. *Instruments* can be controlled by humans as precisely as possible to achieve a controllable performance that human artists desire. They have no autonomy. In contrast to *instruments*, *counterparts* come with some degree of autonomy. They can make decisions and actively contribute to the final result. However, the level at which they operate may differ from case to case. They follow some (usually high-level) instructions from the human co-creator and autonomously create the underlying creative content. One could argue that they exert some level of executive autonomy, while the goal autonomy often primarily is in the hands of the human co-creator. The level on which the systems operate autonomously is in most cases the data. For example, in the visual domain, these data are the pixels, in the audio and music domain they can be notes or audio spectrums, correspondingly in text and speech systems, data might be word tokens.

Goal

Another aspect of co-creation between humans and AI systems is the goal they pursue. Historically, and in terms of typical (non-creative) applications of AI–human interaction, the relationship between autonomously acting AI systems and human users has been the subject of research for many years (Abbass, 2019). In many contemporary use cases, high-level goals are defined by the humans (cf. goal autonomy) as introductions and communicated to the AI. A self-driving car, for example, has the task of driving from one address to another. During the journey, a wide range of decisions must be made, which is usually handled by the AI autonomously (e.g. choosing the route, stopping at red lights, or encountering obstacles on the road). The goal, though, is to get to a specific address given by a human. In a creative environment, however, the goals of humans and AI are different in nature. Unlike the commonly used meaning of goal, human creative intentions cannot be precisely formulated and communicated.

This is, again, related to the nature of creative results, which are beyond predictability and typically ambiguous. Consequently, in an artistic-creative process, it is not purposeful to refer to them as goals. Instead, goals in this context can be seen as artistic introductions of the artist, perhaps with some roughly explicable properties that the artist assumes and must communicate to the system. In a nutshell, the interaction with AI in a creative process is more complex and characterized by an open-ended outcome as compared to other, non-creative AI with predetermined results (e.g. a self-driving car).

Due to the differences in the goal articulation process, it can be argued that the nature of interaction with generative or creative AI appears more likely as true collaboration, when considering the ten challenges coined by Klein et al. (2004) (e.g. through said introductions there might be more focus on "common-grounding activities" and negotiations). Determining an artistic introduction is part of the human input that needs to be communicated with AI in a co-creation process.

As statistical machines, creative AI systems tend to maintain a low-entropy state (see Amini et al., 2023), which can lead to average results that are rather ordinary. This can be countered by stimulating the system to reach new states, for example, by combining different categories or ideas that were not present in the training set, forcing the system to discover new ways to combine these categories. For example, state-of-the-art systems for image generation can be used to create images that combine different categories and concepts that surely have not been existing in the training set (e.g. the concept of a flower with the concept of a chair and make a picture combining different features of those pictures in a reasonable way). In addition, the behaviour of the system can be altered for specific artistic purposes by manually changing the parameters. The main motivation behind this idea is following assumption: an AI system that is able to generate new data must have learned some implicit rules from the domain of the data. For example, a language model has explicitly learned the rules of grammar; a model such as stable diffusion has learned implicit features about numerous concepts that result from the mapping between language and images in its dataset.

Research in the field of explainable AI seeks to formalize and understand these learned characteristics to explain how AI arrives at a particular decision. These features can, at the same time, be a completely new artistic way to explore the artistic space and be explicitly manipulated by artists to create new artistic concepts. This method can be especially useful in the field of contemporary art, where the artworks are highly experimental and work with the most individual concepts and ideas possible. Originality in artistic results can be achieved by changing the learned behaviour of the system to force the system to step out of its average and usual state to achieve new, original results. This can also be done in a collaborative process through various types of communication between human artists and AI. Technical developments in AI will make it increasingly convenient to modify learned parameters of the system for individual purposes. An analogy can be drawn here with the three types of creativity proposed by Margaret Boden, in which exploratory and transformative creativity are explained alongside combinatorial creativity (Boden, 2004). Exploratory creativity can be interpreted as a novel combination of known categories that leads to the exploration of new results from the learned space. In transformative creativity, the conceptual space is reshaped. In this type of creativity, the conceptual space can be interpreted as a transformation of the learned features. However, in her concept, the system transforms its conceptual space autonomously and independently (in order to establish new rules).

Co-Creation Process

Another consequence of the surprising or unexpected nature of creative ideas is that there is not only a single possible output for an artistic introduction of the user. This means that every time the user starts the creative process, the system makes a different suggestion, even if user's introduction remains unchanged. Due to their unpredictable nature, which leads to different results even for the same human introduction, creative AI systems can basically be used repeatedly to make different suggestions for the same request or intention of the human user. It is unpredictable whether the AI's next suggestion will be closer to the human's artistic intent or not. Only the human user can decide when to stop the process of making new suggestions.

In a typical co-creation process, a human artist following a clear artistic intention lets the AI generate different suggestions in each step of the creation, selects one of them, and finally continues the creation process with the next step based on the selection, with or without changing the introduction. This aspect of working with *suggestions* of an artificial counterpart in an artistic scenario did not exist before and is more reminiscent of the way human artists collaborate. This further underscores the unorthodox role of the human artist in a co-creative process with AI. Following the above discussion, it can be said that creative systems, even if they have a degree of (executive) autonomy at the level of underlying low-level data, must be directly or indirectly controlled by human counterparts to fulfil the artistic intentions that only can come from human artists.

However, the aforementioned discussion raises some fundamental questions about the role of humans in an artistic co-creation process with cutting-edge creative AI systems. The intellectual input of humans in such a process is significant to answer questions about the ownership of artworks created with AI systems. The degree of human involvement is currently considered a criterion for whether a song can be awarded a Grammy (Paul, 2023; Miller, 2023) or even be protected by US copyright law (U.S. Copyright Office, 2023). Currently, a composition or an illustration created with an AI system may not be eligible for copyright protection if the level of human involvement is deemed insufficient. However, the current view of how human involvement is estimated seems to focus largely on the underlying data (which is usually produced by AI) and pays less attention to the co-creation process. The previous sections have shown that the assessment of human authorship is not limited to the creation of the underlying data but is instead about human involvement at a higher level in a co-creation process with AI. In these creative scenarios, human involvement can be very deep and involve numerous co-creation steps and refinements. The end result can still be a direct output of AI, as these systems typically operate at the low-level data. On the contrary, since the AI acts autonomously, users can also create results with minimal effort, since the human counterpart is not forced to take further steps but can simply accept AI's first suggestion (but it is also not impossible that a first suggestion exactly fulfils one's artistic intention). These scenarios would be artistically very unclear when it comes to human authorship. The fact is that there is currently no possibility to assess human involvement in a finished work of art.

Future Research Agenda

This chapter aimed to draw attention to the nature of creative AI systems, especially in artistic co-creation scenarios with humans. It indicates that more research is needed to gain a better understanding of the nature of AI creativity in art and the role of humans in such co-creation with AI. The extent of human intervention will certainly vary across art fields and change with new technological developments. In some areas of art (e.g. contemporary art), the distinctiveness of an artwork is particularly important. Artists may use transformative approaches and change the system itself based on their artistic intentions but use the output of the system directly. In this and similar cases, it is not satisfactory to limit authorship to the outcome at the level of the data and pay less attention to the creative process. To conclude the chapter, three questions for future research will be addressed:

RQ1: How can autonomy be defined in an AI-based creative environment?
This question deals with the freedom of AI in an artistic co-creation process with humans. It describes the possible levels in which AI can act autonomously.

RQ2: What is the role of humans in a co-creation process in a creative environment?

The artistic role of humans in a co-creative process can be conceptualized in a wide range of possibilities. Artistic research is needed to define a conceptual space for these possibilities. Additionally, further research is required to find technical solutions for implementing the artistic requirements that arise from that conceptual space. This includes, for example, communication concepts between AI and human counterparts.

RQ3: How do the responses to the above questions affect authorship in this context?

Human contribution to an artistic work is not only crucial for answering questions of authorship posed in the field of law, but it also matters for artistic practice. The answer to this question relies on the conceptual space examined in the first and second questions, in which the role of the human artists is to be examined.

These questions not only depend on the current state of the technology and its limitations, but also affect any human AI co-creation process in the field of art in a broader sense. They also aim to draw attention to a clearer understanding of the nature of AI creativity in art and the role of humans in such co-creation process.

Acknowledgements

Karin Breckner's contribution to this chapter has been conducted within the scope of the Human-Centered Artificial Intelligence (HCAI) project, funded by the Austrian Science Fund (FWF) [DFH 23-N]. Thomas Neumayr, Frederik Hirschmann, and Mirjam Augstein's work on this chapter has been conducted within the scope of the Hybrid Collaboration Spaces (HYCOS) project, also funded by the FWF [P 34928].

References

Abbass, H. (2019). Social integration of artificial intelligence: Functions, automation allocation logic and human-autonomy trust. *Cognitive Computation*, *11*. https://doi.org/10.1007/s12559-018-9619-0

Agostinelli, A., Denk, T. I., Borsos, Z., Engel, J., Verzetti, M., Caillon, A., Huang, Q., Jansen, A., Roberts, A., Tagliasacchi, M., Sharifi, M., Zeghidour, N., & Frank, C. (2023). *MusicLM: Generating music from text*. https://arxiv.org/abs/2301.11325

Amini, A., Cotterell, R., Hewitt, J., Meister, C., & Pimentel, T. (2023). Generating text from language models. In *Proceedings of the 61st annual meeting of the association for computational linguistics (vol. 6: Tutorial abstracts)* (pp. 27–31). Association for Computational Linguistics. https://aclanthology.org/2023.acl-tutorials.4, https://doi.org/10.18653/v1/2023.acl-tutorials.4

Becker, H. (1982). *Art worlds*. University of California Press.

Boden, M. A. (2004). *The creative mind: Myths and mechanisms*. Routledge.

Boden, M. A. (2007). Creativity in a nutshell. *Think*, *5*(15), 83–96. https://doi.org/10.1017/S147717560000230X
Calvo, R. A., Peters, D., Vold, K., & Ryan, R. M. (2020). Supporting human autonomy in ai systems: A framework for ethical enquiry. In *Ethics of digital well-being: A multidisciplinary approach* (pp. 31–54). Springer.
Castelfranchi, C. (1994). Guarantees for autonomy in cognitive agent architecture. In *International workshop on agent theories, architectures, and languages* (pp. 56–70). Springer.
Cetinic, E., & She, J. (2022). Understanding and creating art with AI: Review and outlook. *ACM Transactions on Multimedia Computing, Communications, and Applications (TOMM)*, *18*(2), 1–22.
ChatGPT, OpenAI. (2022). *Introducing ChatGPT*. Retrieved November 27, 2023, from https://openai.com/blog/chatgpt
Csikszentmihalyi, M. (1997). *Creativity: Flow and the psychology of discovery and invention*. HarperCollins Publishers.
Denning, P. J., & Yaholkovsky, P. (2008). Getting to "we". *Communications of the ACM*, *51*(4), 19–24.
Dhariwal, P., Jun, H., Payne, C., Kim, J. W., Radford, A., & Sutskever, I. (2020). *Jukebox: A generative model for music*. https://arxiv.org/abs/2005.00341
Goodfellow, I., Pouget-Abadie, J., Mirza, M., Xu, B., Warde-Farley, D., Ozair, S., Courville, A., & Bengio, Y. (2014). Generative adversarial networks. *Advances in Neural Information Processing Systems*, *3*. https://doi.org/10.1145/3422622
Gownder, J. P., O'Grady, M., McQuivey, J., Curran, R., McElveen, S., Suresh, A., & Starks, D. (2023). *Forrester's 2023 generative AI jobs impact forecast, US*. Retrieved November 19, 2023, from https://regmedia.co.uk/2023/09/06/forrester_gen_ai_report_pdf
Heinich, N. (2003). The Van Gogh effect. In *The sociology of art: A reader* (pp. 122–131). Routledge.
Hieslmair, M. (2002). *Holly+: Come and sing like Holly Herndon!* Retrieved November 15, 2023, from https://ars.electronica.art/aeblog/en/2022/06/15/hollyplus/
Jennings, K. E. (2010). Developing creativity: Artificial barriers in artificial intelligence. *Minds and Machines*, *20*, 489–501.
Jucker, J.-L., Barrett, J. L., & Wlodarski, R. (2014). "I just don't get it": Perceived artists' intentions affect art evaluations. *Empirical Studies of the Arts*, *32*(2), 149–182.
Klein, G., Woods, D. D., Bradshaw, J. M., Hoffman, R. R., & Feltovich, P. J. (2004). Ten challenges for making automation a "team player" in joint human-agent activity. *IEEE Intelligent Systems*, *19*(6), 91–95.
LeCun, Y., Bengio, Y., & Hinton, G. (2015, May). Deep learning. *Nature*, *521*, 436–444. https://doi.org/10.1038/nature14539
London, S. (1995). *Collaboration and community* (pp. 5–6). Pew Partnership for Civic Change, University of Richmond.
McKenzie, T. (2022). *The crow: An AI film that won Cannes short film festival*. Retrieved November 15, 2023, from https://80.lv/articles/the-crow-an-ai-film-that-won-cannes-short-film-festival
Miller, M. (2023). *AI and copyright: How the recording academy is leading the conversation to protect music creators*. Retrieved September 6, 2023, from www.grammy.com/news/ai-copyright-protecting-music-creators-united-states
Nikrang, A. (2020). *Ricercar, an AI-based musical companion*. Retrieved August 13, 2023, from https://ars.electronica.art/futurelab/en/fie-ricercar/
Paul, L. (2023). *Music made with AI will be eligible for Grammys, but only humans will be awarded*. Retrieved September 6, 2023, from www.rollingstone.com/music/music-news/ai-created-music-with-human-contrib

Payne, C. (2019, April 25). Payne, Christine "MuseNet". *OpenAI*. Retrieved August 13, 2023, from openai.com/blog/musenet
Radford, A., Narasimhan, K., Salimans, T., & Sutskever, I. (2018). *Improving language understanding by generative pre-training*.
Shah, C. (2010). Collaborative information seeking: A literature review. *Advances in Librarianship*, 3–33.
Sohl-Dickstein, J., Weiss, E., Maheswaranathan, N., & Ganguli, S. (2015, June). Deep unsupervised learning using nonequilibrium thermodynamics. In *International conference on machine learning* (pp. 2256–2265). PMLR.
Tarr, B., Launay, J., & Dunbar, R. (2014, 09). Music and social bonding: "Self-other" merging and neurohormonal mechanisms. *Frontiers in Psychology*, *5*, 1096. https://doi.org/10.3389/fpsyg.2014.01096
Unsupervised. (2022). Refik Anadol exhibition. *MoMa*. Retrieved November 15, 2023, from www.moma.org/calendar/exhibitions/5535
U.S. Copyright Office, L. o. C. (2023). *Copyright registration guidance: Works containing material generated by artificial intelligence*. Retrieved September 6, 2023, from www.federalregister.gov/d/2023-05321
Vaswani, A., Shazeer, N., Parmar, N., Uszkoreit, J., Jones, L., Gomez, A. N., Kaiser, Ł., & Polosukhin, I. (2017). Attention is all you need. *Advances in Neural Information Processing Systems*, 5998–6008.
Wingstroem, R., Hautala, J., & Lundman, R. (2022, August). Redefining creativity in the era of AI? Perspectives of computer scientists and new media artists. *Creativity Research Journal*, 1–17. https://doi.org/10.1080/10400419.2022.2107850
Zhao, W. X., Zhou, K., Li, J., Tang, T., Wang, X., Hou, Y., Min, Y., Zhang, B., Zhang, J., Dong, Z., Du, Y., Yang, C., Chen, Y., Chen, Z., Jiang, J., Ren, R., Li, Y., Tang, X., Liu, Z., . . . Wen, J. (2023). *A survey of large language models*. *ArXiv*. https://arxiv.org/abs/2303.18223.

Chapter 2

Creativity as a Socio-Technical Phenomenon

Implication for the Research on AI and Creativity

Mariya Dzhimova

Introduction

Over the past decades, creativity has been seen as a driver of both personal fulfilment and economic growth. Recently, with the rise of artificial intelligence (AI), and in particular generative AI, the notion of creativity is experiencing a new upswing, as there is a lively discussion about how AI will extend, accelerate, or possibly even replace human creativity. What is striking about this discussion is that creativity is often equated with artistic creativity, or even with the concept of art. Perhaps this is due to the fact that art has long been seen as the epitome of creativity (Reckwitz, 2008). At the heart of this assumption, which has its historical origins in European Romanticism, is an irreplaceable human individual, a "creator", a genius endowed with extraordinary mental traits that enable the individual to produce aesthetic works that break with existing (aesthetic) rules and are thus in this sense novel, original and creative. Since this Romantic approach differs from the previous period of imitative aesthetics, which was concerned with perfecting the rules and reproducing the old (Reckwitz, 2012), it also assumes that the artist does not follow a certain rationality and certain rules but is driven by intuition and emotion (Manovich, 2022). Due to this intertwined relationship with the Romantic notion of art and the figure of the artist, creativity has often been understood as aesthetic expression and as thus as the production of novel artworks, which in turn represent a break with the (aesthetic) old and are the result of an extraordinary, almost mysterious action that does not follow any particular rationality. This understanding of creativity seems to be reflected in the AI debate, as it steers the discussion in a certain direction: (1) creativity is often seen as an aesthetic phenomenon, and the creativity of the machine is measured by its ability to create novel and original works of art (Manovich, 2022); (2) often, it is argued that AI is not capable of creating something new that breaks with the old, because AI has been trained on the old; (3) creativity is often seen as a psychological phenomenon, and so the discussion revolves around the importance of certain human cognitive and personal traits for creativity and the extent to which AI possesses these traits

DOI: 10.4324/9781003453901-4

or not. Furthermore, as a result of this psychologization of creativity, it is often assumed that only humans can act as autonomous creators, while all technologies and materials involved in production are merely tools with the help of which the ideas that already exist virtually in the creator's head are merely realized. It is therefore often argued that AI will change or is already changing creative practice by leading to new collaboration between human and machine. This narrowing of the discussion about AI and creativity is problematic for several reasons.

First, there are other social fields outside of art that understand, describe and reflect themselves as creative (Hutter et al., 2016). In recent decades, "being creative" and "being innovative" have been seen as central norms of action in the economy, in general, and in the so-called creative industries, in particular (Jones et al., 2015), where creativity is seen as both the most important competitive factor and the ultimate goal. However, not only the economy but also other social spaces are permeated by the "logic of the total creative imperative" (Raunig & Wuggenig, 2016, p. 71) or the "creativity dispositive" (Reckwitz, 2012). This means that "being creative" is no longer an option but a deeply rooted desire and imperative that applies not only to companies but also to entire cities (Landry, 2012), social classes (Florida, 2002), milieus (Merkel, 2008) and private lifestyles (Reckwitz, 2012). Thus, the idea of creativity has established itself as the guiding category of modernity (Göttlich & Kurt, 2012), and as such it can be described as a main characteristic of contemporary societies.

On the other hand, while the process of aestheticization has extended into many areas of society, actors in the arts have responded by rejecting creativity (Van den Berg, 2009). As Van den Berg (2009) points out, many artists felt challenged to take a critical stance on the imperative of creativity and to reflect on what role art can play in an environment where creativity is ubiquitous. Accordingly, in the very place that is considered the epitome of creativity, namely the arts, one finds approaches and works that have critically distanced themselves from the concepts inherent to creativity, such as "authorship", "subjectivity" and "novelty".

Second, creativity can be understood not only as a specialized activity carried out in specific social fields dedicated to the production of the new, such as art, science and economics but also as a basic competence of human action in general (Joas, 1996). In this second anthropological sense, creativity is understood much more broadly, as the entire spectrum of human action – from conscious reflection to the ability to solve problems through situational experimentation and deviation from the rules and routine, and thus as the basis for societal change (Schäfer, 2012). However, deviant behaviour and thus the production of novelty does not occur *ex nihilo* but, as Schäfer (2012) has pointed out, stands in a relationship of continuity and tension with routine or with habitus as an integrated ensemble of "schemes of perception, thought and action" (Bourdieu, 1987, p. 101).

Finally, creativity can be seen as a social rather than a psychological phenomenon, and thus as a result of social interactions rather than specific cognitive and personality traits. This means that the production of creative ideas, the perception and evaluation of these ideas as creative and their implementation in the relevant domain are collective achievements and social processes. Moreover, the sociality of creativity is not only limited to human interactions but also between humans and technologies. In this respect, creativity can be considered as the result of collaboration between human and non-humans (technologies, tools, materiality, etc.) and thus as a socio-technical phenomenon.

Taking into account the societal, anthropological and socio-technical dimensions of creativity can greatly broaden the view and discussion on AI and creativity. However, it is beyond the scope of this chapter to consider all of these arguments in detail. The focus of this chapter is therefore on the conceptualization of creativity as a socio-technical phenomenon and its implication for the research on AI and creativity. This chapter is organized into two sections. The first section presents a brief overview of different theoretical approaches that have examined creativity as a collective achievement. The various studies are presented along the processes of production, evaluation and adoption of creative ideas. Based on these findings, the second section formulates questions relevant to research on AI and creativity.

Creativity as a Socio-Technical Phenomenon

In the early days of creativity research, from the mid-1950s to the mid-1970s, creativity was seen as a psychological phenomenon and was primarily the subject of psychological research and related cognitive sciences (Nyfeler, 2019). In this context, the central research question is, "What is it that makes someone creative?" (Weisberg, 2006, p. 596). Accordingly, the scientific focus at that time was on the individual and his or her personality traits, which were seen as the cause of creative abilities. This changed in the following decades, when the focus of creativity research shifted from a cognitivist and personality-centred psychological approach to a social science approach that views creativity as a social phenomenon (Nyfeler, 2019), that is emphasizing the social embedding and conditionality of the production of creative ideas as well as their evaluation and adoption.

The Production Processes

Creativity as Collective Achievement

That creativity is a social phenomenon means that it is the result of a complex and dynamic interaction of actors, who in turn are embedded in a social context (Paulus & Nijstad, 2019). The social context is conceived in different ways depending on the research interest, namely as a particular organizational or institutional environment, as a social network, as a sociocultural framework, as a physical-material setting and so on, or even as a nation and whole civilization (Simonton, 2019). Especially research within social psychology, organization and management studies, as well as innovation studies illustrated that creativity is the outcome of interpersonal relationships and interactions, collective action as well as a specific group and team constellations (Amabile, 1983; Kurtzberg & Amabile, 2010; Lee et al., 2015) and that particular contextual factors can generate, promote or also inhibit creativity (Amabile, 1996, 1998; Andriopoulos, 2001; Perry-Smith, 2006; Watson, 2007; Rosso, 2014; Sonenshein, 2016). For example, Perry-Smith and Shalley (2003) examine the dynamic and complex interplay between certain social network features such as strength of the ties and networks' position, on the one hand, and creative performance in organizations, on the other hand. They argue that both a large number of weak ties and peripheral positions in a given network foster creativity in organizations: (1) weaker ties provide access to a wider range of people and more non-redundant information and (2) peripheral position with external and boundary-spanning connections provide actors with approaches and perspectives that are used in relevant but different fields. However, they pointed out that this relationship is not linear, as a very large number of weak ties may become distracting and constrain creativity. Further, the relationship between network position and creativity is moderated by contextual factors such as diversity relative to the network, cultural norms and the tightness of the symbolic structure of the dominant field.

Other studies go a step further such as the work of Fortwengel et al. (2016) who point out that the social context, usually defined as the organizational environment, often serves as an explanation for creativity, while creativity is primarily seen as a dependent and static outcome. Following Giddens's (1984) structuration theory, which assumes that social structures are constituted by action, but at the same time serve as a medium of it, Fortwengel et al. (2016) conceptualize creativity less as a statistical *outcome*, but rather as a dynamic *process* that is continuously organized. In this vein, Drazin et al. (1999) and Hargadon and Bechky (2006) have also emphasized the temporality of creativity as well as that creativity is the result of both structures and actions interacting in complex ways over time and at multiple levels. For example, Sonenshein (2014, 2016) have shown how various contextual factors, such as resource endowments and routines, are dynamically enacted and thus both enable and constrain the production of creative outcomes over time.

Besides organization and management studies, research in the sociology of art has also contributed to the conceptualization of creativity as a social phenomenon. Pierre Bourdieu's (1999) approach to the cultural field and Howard S. Becker's (2008) approach to art worlds are particularly noteworthy. Although neither theory explicitly addresses the phenomenon of creativity, their approaches are important, in that artistic and cultural production is regarded as the result of collective action (Becker, 2008) and negotiations about what is or is not considered legitimate art (Bourdieu, 1999). According to Becker (2008), art is not the work of a single ingenious creator, but the result of the cooperation of many social actors within the art world, such as the artists, but also the support personnel who provide various resources within the production process, the various instances of art mediation and art criticism and finally the audience. This cooperation, in turn, is characterized by a reliance on shared and established conventions about how art is to be produced, what counts as art and what does not and how art is to be understood and experienced emotionally and aesthetically. Similarly, in Pierre Bourdieu's theory, art objects are also effects of the social dynamics and logic of the art field, which in turn is based on power struggles over the attribution of legitimate art. Different actors within the art field embody different aesthetic conceptions and struggle with each other to assert and establish their own aesthetics as legitimate. The different aesthetic conceptions in turn depend on the respective accumulation of capital and the associated positioning of the actors in the social space and are negotiated and stabilized through everyday practice (Bourdieu, 1999).

The Material Dimension of the Creative Process

Later research in the sociology of art has extended this more anthropocentric view of creativity by drawing on theoretical developments in Science and Technology Studies, and in particular Actor-Network Theory (ANT) (Latour, 2005; Law & Hassard, 1999). According to this approach, artworks and creativity are the result of a process that unfolds through the mutual performativity and interaction of humans as well as non-humans such as objects, instruments, systems and technologies. Thus, the agency of materiality and the engagement with it has always been crucial to the process by which artworks are created (Rubio, 2012; Yaneva, 2003; Schürkmann, 2018). The mediation (ANT), resistance (Pickering, 1995) and affordance (Gibson, 1986) of materiality and engagement with it are fundamental to practice, in general, and creative practice, in particular. And this applies not only to the use of high technologies such as AI but also to all materiality, including oil paint (Lehmann, 2012) or Degas's electric pencil ("crayon électrique") (Berger, 2012), to name just two examples. In this vein, Rubio (2012) empirically investigates the production of the Spiral Jetty, an earthwork sculpture created by the American artist Robert Smithson by taking into account the agency of the materiality. Rubio (2012) shows in his study that the concrete form of the artwork was not a simple and direct mechanical realization of an abstract idea in the artist's mind, but that the artwork unfolded step by step as a result of the various practical problems and constraints, but also of the possibilities and solutions that arose from working with certain technologies and materials. So, the site and the materiality functioned as "affordances" in the sense of Gibson (1986), which made certain ways of thinking and acting possible. Accordingly, the artistic creative process is defined as distributed between humans and non-humans.

The Evaluation and Adoption Processes of Creativity

Perception, Evaluation and Legitimation of Creativity

The sociality of creativity lies not only in the fact that creativity is the result of a complex interplay of collective actions and a specific socio-technical context but also in the fact that creativity is defined as a social construct. This means that creativity is not a straightforward fact, but it first has to be "fabricated" (Nyfeler, 2019), "invented" (Reckwitz, 2012), made into such through practices of perception, evaluation and legitimation. Accordingly, ideas, objects, processes, practices and the like are not creative per se but must be recognized and interpreted as *new* as well as evaluated as *better* or more *valuable*, and accordingly labelled as creative (Amabile, 1983; Luhmann, 1988; Boden, 2004). These processes of interpretation and evaluation are also social processes carried out by relevant group, social field, broader institutional context or society. Mihaly Csikszentmihalyi is one of the first cultural anthropologists to conceive of creativity as a social construct, arguing "who brings about some change in the field, a change that the field will consider to be creative" (Csíkszentmihályi, 1988, p. 325).

Further, these social processes of interpretation and evaluation are distributed among a multitude of heterogeneous stakeholders (experts) who base their interpretations and evaluations on certain conventions, rules and values in the respective social field (Becker, 2008; Bourdieu, 1999). As Reckwitz (2012, p. 44) points out, "it depends on social criteria that say which new is valuable and which is not". These criteria vary depending on the different areas of society, such as art, science and business (Rammert, 2010). Further, the heterogeneous stakeholders who also figure as gatekeepers have different power resources to legitimize something as creative (Bourdieu, 1999) and form strategic networks with each other in order to institutionally distribute and establish creative ideas, products, processes and the like in the relevant social field or in society (Hutter et al., 2016; Valente, 1996).

What creative is, is subject to a process of legitimation and "something that is not legitimate or accepted cannot be creative" (Cattani & Ferriani, 2008, p. 826). In addition, as the attribution and legitimation of creativity is controversial and contested, it must also be constantly renegotiated (Bourdieu, 1999). However, as Alexander and Bowler (2021, p. 14) pointed out, "legitimacy is achieved, even in the face of continuing contestation, as indeed legitimizing is a continual, never-ending process".

Adoption of Creative Ideas

Further, to make contributions to any domain or social field requires not only creative ideas be generated as well as recognized and valued as creative but also that these ideas are adopted by relevant stakeholders and implemented in the relevant domain (Csikszentmihalyi, 1999). By reviewing some of the theoretical frameworks to innovation adoption, both Passoth (2012) and Wisdom et al. (2014) have summarized that the implementation of innovations is a complex process that takes place over different non-linear time stages and at multi-levels. For example, according to Geels's (2004) theoretical framework, these levels are the micro-level of niches, the meso-level of socio-technical regimes and the macro-level of socio-technical landscape developments. The socio-technical regime includes stable and established institutional structures of governments, economy, science and infrastructures, which change slowly and define a framework that maintains the status quo and allows the scaling and spread of new ideas and approaches, whereas the landscape developments include mega-trend such as climate change. New and alternative ideas and approaches initially spread in niches, but when tensions and mismatches occur, "windows of opportunity" are created and new and creative ideas can take advantage and breakthrough in the relevant domain. Eventually a new system and regime emerges, supported by a network of social groups that create and maintain socio-technical systems. The new system may eventually influence wider landscape developments. Furthermore, as Wisdom et al. (2014) pointed out, the adoption process is influenced by various external factors such as the sociocultural context, sociopolitical environment, government policies and regulations, as well as internal factors such as organizational (norms, values, culture, size, structure, etc.) and innovation characteristics (complexity, relative advantage and observability, etc.), to name but a few.

It can be summarized that the production as well as the evaluation and adoption of creative ideas are socially conditioned and embedded processes, whereby all three phases are interrelated and recursive: for example, the evaluation criteria of the respective social field already play a role in the creative production process insofar as the respective idea, process or product seeks social recognition and thus does not represent a complete break with the established rules (Groys, 1992; Becker, 2008; Reckwitz, 2012).

Future Research Agenda

Despite the many meanings attributed to the concept of creativity in contemporary societies, the view that creativity is a psychological phenomenon still dominates, and the question of specific abilities and traits remains central. This view also seems to shape the debate on AI and creativity. However, if we take into account social science research, and thus the assumption that creativity is a social phenomenon, new and exciting questions arise about the relationship between AI and creativity. Some of these questions are raised below.

Although, as noted above, the production, evaluation and adoption of creative ideas are intertwined, here the research questions are formulated along the three distinct stages.

The Production Process of Creative Outcomes

The production of creative outcomes unfolds in a complex and dynamic interaction between humans and non-humans, embedded in an institutionally and culturally variable context and takes place over time. In this respect, creativity is a collective achievement and thus the product of a process of co-creation long before the use of AI tools. However, considering that AI tools derive their own, often complex, heuristics, which are often incomprehensible, and thus produce unpredictable and surprising results, AI technologies might go far beyond the agency of other objects, tools and digital technologies in terms of altered efficacy (ANT), resistance (Pickering, 1995) and affordances (Gibson, 1986). Thus, the autonomy and learning capacity of both humans and non-humans, and the resulting feedback loops, can shape the interaction between artist and machine in novel ways, opening up new dynamics and forms of collaboration and distribution of creativity. If this is the case, however, needs to be examined empirically in the respective concrete practice, which is embedded in a specific socio-technical application field. Against this background, the following research question arises:

RQ1: Does the use of AI lead to qualitatively new forms of co-creation, or does it merely represent an acceleration of the production process of creative outcome?

Identification and Evaluation Processes of Creative Outcomes

An idea, an object, a process is not creative per se but must be defined and evaluated as new and valuable by the relevant actors in the respective social field according to certain criteria and rules in order to be considered creative. Thus, creativity is a historically variable social construct. Therefore, it seems important to empirically investigate contemporary discursive practices on AI and creativity to understand whether new interpretations and explanatory models of creativity are being demanded and designed accordingly. In the debate on AI and creativity, it is often said that AI will lead to a paradigm shift in creativity (Alexeev, 2023) or that a new creative paradigm is needed (Tigre Moura, 2023). So, it is possible that the use of AI tools in creative practice could lead to a social change in the regimes of justification. Against this background, the following question is relevant:

RQ2: What rules and values guide the actors in the respective social field when they describe AI-generated output as creative?

Adoption Process of Creative Outcomes

Ultimately, these ideas, products, processes and the like go down in history as creative achievements that have spread and become established and accepted in their respective social fields, thereby changing social practice. The social diffusion and establishment of new ideas are complex and multi-stage and multi-level social processes influenced by multiple factors (Wisdom et al., 2014) and distributed among many heterogeneous actors (Rammert, 2010; Becker, 2008). These stakeholders in turn pursue different interests, have different power resources and form strategic networks in order to establish something that is perceived as new. In addition, as the Social Construction of Technology (Bijker et al., 2012) approaches have shown, the social integration of a new technology depends not only on its technical performance but also on social factors. This also applies to generative AI technologies and their use in the creation process. Although many generative AI tools such as ChatGPT have spread very quickly among users, their permanent adoption in the respective social practices is still open. In particular, potential risks of generative AI, such as incorrect results, so-called hallucinations, the tendency of generative models to go MAD (Model Autophagy Disorder) (Alemohammad et al., 2023), misinformation, biases in the training data, competition, regulatory attempts at national and international level to control these risks, as well as power games and negotiations between the various AI companies, might have a major impact on the diffusion and adoption of the use of AI tools, in general, and on AI-generated content, in specific. Against this background, the questions arise:

RQ3: What mechanisms of diffusion and adoption of various AI tools can be observed in the respective creative process in the respective social field? What are the internal and external factors that could affect the adoption process? Who are the relevant stakeholders, who has joined them; what interests are being pursued; and what power resources do they have at their disposal to enforce the use of generative AI in the production process of creative outcome?

References

Alemohammad, S., Casco-Rodriguez, J., Luzi, L., Humayun, A. I., Babaei, H., LeJeune, D., Siahkoohi, A., & Baraniuk, R. G. (2023). *Self-consuming generative models go mad. arXiv Preprint.* https://arxiv.org/abs/2307.01850

Alexander, V. D., & Bowler, A. E. (2021). Contestation in aesthetic fields: Legitimation and legitimacy struggles in outsider art. *Poetics*, *84*, 101485.

Alexeev, V. (2023). *Wir sind am Anfang einer neuen Kulturepoche*. www.zeit.de/kultur/2023-09/vladimir-alexeev-kuenstliche-intelligenz-kunst-merzmensch
Amabile, T. M. (1983). The social psychology of creativity: A componential conceptualization. *Journal of Personality and Social Psychology*, *45*(2), 357–376.
Amabile, T. M. (1996). *Creativity in context*. Westview Press.
Amabile, T. M. (1998). How to kill creativity. *Harward Business Review*, *76*(5), 76–87.
Andriopoulos, C. (2001). Determinants of organisational creativity: A literature review. *Management Decision*, *39*(10), 834–841.
Becker, H. S. (2008). *Art worlds*. University of California Press.
Berger, C. (2012). Multiplikation und Diversifikation der Bilder und der Akteure in Edgar Degas' künstlerischer Praxis. *Zeitschrift für Ästhetik und Allgemeine Kunstwissenschaft*, *57*(1), 90–103.
Bijker, W. E., Hughes, T. P., & Pinch, T. (2012). *The social construction of technological systems* (Anniversary ed.). MIT Press.
Boden, M. A. (2004). *The creative mind: Myths and mechanisms*. Psychology Press.
Bourdieu, P. (1987). *Sozialer Sinn. Kritik der theoretischen Vernunft*. Suhrkamp.
Bourdieu, P. (1999). *Die Regeln der Kunst: Genese und Struktur des literarischen Feldes*. Suhrkamp.
Cattani, G., & Ferriani, S. (2008). A core/periphery perspective on individual creative performance: Social networks and cinematic achievements in the Hollywood film industry. *Organization Science*, *19*(6), 824–844.
Csíkszentmihályi, M. (1988). Society, culture, and person: A system view of creativity. In R. J. Sternberg (Eds.), *The nature of creativity* (pp. 325–339). Cambridge University Press.
Csikszentmihalyi, M. (1999). Implications of a systems perspective for the study of creativity. In R. J. Sternberg (Eds.), *Handbook of creativity* (pp. 313–335). Cambridge University Press.
Drazin, R., Glynn, M. A., & Kazanjian, R. K. (1999). Multilevel theorizing about creativity in organizations: A sensemaking perspective. *Academy of Management Review*, 24(2), 286–307.
Florida, R. (2002). *The rise of the creative class*. Basic Books.
Fortwengel, J., Schüssler, E., & Sydow, J. (2016). Studying organizational creativity as process: Fluidity or duality? *Creativity and Innovation Management*, 2(2), 1–12.
Geels, F. W. (2004). From sectoral systems of innovation to socio-technical systems: Insights about dynamics and change from sociology and institutional theory. *Research Policy*, *33*(6–7), 897–920.
Gibson, J. J. (1986). *The ecological approach to visual perception*. Psychology Press.
Giddens, A. (1984). *The constitution of society: Outline of the theory of structuration*. University of California Press.
Göttlich, U., & Kurt, R. (2012). *Kreativität und Improvisation: Soziologische Positionen*. Springer.
Groys, B. (1992). *Über das Neue: Versuch einer Kulturökonomie*. Hanser.
Hargadon, A. B., & Bechky, B. A. (2006). When collections of creatives become creative collectives: A field study of problem solving at work. *Organization Science*, *17*(4), 484–500.
Hutter, M., Knoblauch, H., Rammert, W., & Windeler, A. (2016). Innovationsgesellschaft heute: Die reflexive Herstellung des Neuen. In *Innovationsgesellschaft heute: Perspektiven, Felder und Fälle* (pp. 15–35). Springer.
Joas, H. (1996). *Die Kreativität des Handelns*. Suhrkamp.
Jones, C., Lorenzen, M., & Sapsed, J. (Eds.). (2015). *The Oxford handbook of creative industries*. Oxford University Press.
Kurtzberg, T. R., & Amabile, T. M. (2010). From Guilford to creative synergy: Opening the black box of team-level creativity. *Creativity Research Journal*, *13*(3), 285–294.

Landry, C. (2012). *The creative city: A toolkit for urban innovators*. Earthscan.

Latour, B. (2005). *Reassembling the social: An introduction to actor-network-theory*. Oxford University Press.

Law, J., & Hassard, J. (1999). *Actor network theory and after*. Wiley-Blackwell.

Lee, Y.-N., Walsh, J. P., & Wang, J. (2015). Creativity in scientific teams: Unpacking novelty and impact. *Research Policy*, *44*(3), 684–697.

Lehmann, A.-S. (2012). Das Medium als Mediator Eine Materialtheorie für (Öl-) Bilder. *Zeitschrift für Ästhetik und allgemeine Kunstwissenschaft*, *57*(1), 69–88.

Luhmann, N. (1988). Über Kreativität. In H.-U. Gumbrecht (Ed.), *Kreativität – Ein verbrauchter Begriff?* (pp. 11–19). Wilhelm Fink Verlag.

Manovich, L. (2022). AI and myths of creativity. *Architectural Design*, *92*(3), 60–65.

Merkel, J. (2008). *Kreativquartiere. Urbane Milieus zwischen Inspiration und Prekariat*. Edition Sigma.

Nyfeler, J. (2019). *Die Fabrikation von Kreativität: Organisation und Kommunikation in der Mode*. Bielefeld University Press.

Passoth, J. H. (2012). Heterogene Praktiken, variable Kreativitäten. In U. Göttlich & R. Kurt (Eds.), *Kreativität und Improvisation: Soziologische Positionen* (pp. 45–62). Springer.

Paulus, P. B., & Nijstad, B. A. (Eds.). (2019). *The Oxford handbook of group creativity and innovation*. Oxford Library of Psychology.

Perry-Smith, J. E. (2006). Social yet creative: The role of social relationships in facilitating individual creativity. *Academy of Management Journal*, *49*(1), 85–101.

Perry-Smith, J. E., & Shalley, C. E. (2003). The social side of creativity: A static and dynamic social network perspective. *Academy of Management Review*, *28*, 89–106.

Pickering, A. (1995). *The mangle of practice: Time, agency, and science*. University of Chicago Press.

Rammert, W. (2010). Die Innovationen der Gesellschaft. In J. Howaldt & H. Jacobsen (Eds.), *Soziale Innovation: Auf dem Weg zu einem postindustriellen Innovationsparadigma* (pp. 21–51). VS Verlag.

Raunig, G., & Wuggenig, U. (Eds.). (2016). *Kritik der Kreativität*. Turia+ Kant.

Reckwitz, A. (2008). *Unscharfe Grenzen. Perspektiven der Kultursoziologie*. Transcript.

Reckwitz, A. (2012). *Die Erfindung der Kreativität: Zum Prozess gesellschaftlicher Ästhetisierung*. Suhrkamp.

Rosso, B. D. (2014). Creativity and constraints: Exploring the role of constraints in the creative processes of research and development teams. *Organization Studies*, *35*(4), 551–585.

Rubio, F. D. (2012). The material production of the Spiral Jetty: A study of culture in the making. *Cultural Sociology*, *6*(2), 143–161.

Schäfer, H. (2012). Kreativität und Gewohnheit. Ein Vergleich zwischen Praxistheorie und Pragmatismus. In U. Göttlich & R. Kurt (Eds.), *Kreativität und Improvisation: Soziologische Positionen* (pp. 17–43). Springer.

Schürkmann, C. (2018). Über das Sichtbare hinaus: Eine Soziologie künstlerischer Praxis. *Zeitschrift für Soziologie*, *47*(6), 438–453.

Simonton, D. K. (2019). Creativity in sociocultural systems. In P. B. Paulus & B. A. Nijstad (Eds.), *The Oxford handbook of group creativity and innovation* (pp. 270–284). Oxford Library of Psychology.

Sonenshein, S. (2014). How organizations foster the creative use of resources. *Academy of Management Journal*, *57*(3), 814–848.

Sonenshein, S. (2016). Routines and creativity: From dualism to duality. *Organization Science*, *27*, 739–758.

Tigre Moura, F. (2023). Artificial intelligence, creativity, and intentionality: The need for a paradigm shift. *The Journal of Creative Behavior*, *57*(3), 336–338.

Valente, T. W. (1996). Social network thresholds in the diffusion of innovations. *Social Networks*, *18*(1), 69–89.

Van den Berg, K. (2009). Drei Absagen der Kunst an ihren erweiterten Begriff. In A. J. Stephan, E. Schröter, & N. Stehr (Eds.), *Rationalität der Kreativität? Multidisziplinäre Beiträge zur Analyse der Produktion, Organisation und Bildung von Kreativität* (pp. 207–224). VS Verlag.

Watson, E. (2007). Who or what creates? A conceptual framework for social creativity. *Human Resource Development Review*, *6*(4), 419–441.

Weisberg, R. W. (2006). *Creativity: Understanding innovation in problem solving, science, invention, and the arts*. John Wiley & Sons.

Wisdom, J. P., Chor, K. H. B., Hoagwood, K. E., & Horwitz, S. M. (2014). Innovation adoption: A review of theories and constructs. *Administration and Policy in Mental Health and Mental Health Services Research*, *41*, 480–502.

Yaneva, A. (2003). Chalk steps on the museum floor. *Journal of Material Culture*, *8*(2), 169–188.

Chapter 3

Rethinking Creativity Frameworks for Artificial Intelligence

Francisco Tigre Moura

Introduction

The developments and fast adoption of artificial intelligence (AI) tools for creative tasks has sparked a profound transformative era, reshaping the landscape of numerous industries, including the creative sector. In such realms, the perceptions of creative processes and the value of creative outputs, for example, have been deeply rooted in notions of subjectivity, emotion, skill and expertise, and human intentionality. The current creativity frameworks (e.g. 4Ps, 5As, 7Cs, and 8Ps), extremely relevant for the understanding of creative theories and concepts, are fundamentally grounded on such assumptions and have successfully provided the basis for the existing creativity paradigm.

However, the past decade has witnessed the exponential development and adoption of intelligent system applications for creative processes. They have become highly sophisticated, capable of performing creative tasks which require both rational and abstract thinking abilities, such as writing poetry, composing music, and designing visual art. Importantly, such outputs have shown to rival human-made ones (without the support of AI) and equally capable of eliciting different forms of positive responses in humans, including affective ones.

Furthermore, the various forms and levels of artificial co-creation (collaboration between humans and AI), fundamentally alter the expected role and influence of humans during creative processes. For example, AI acts as an external creative augmentation agent, capable of enhancing human creative potential and skill. Also, it democratizes the necessary creative expertise for various creative processes, enabling, for example, the creativity of "non-creatives". In contrast, the lack of need for creative expertise may negatively impact the self-identity of human creatives. Furthermore, the abundance of AI-generated, or co-created, content may potentially affection the perception of the public to an extent of devaluating human creative skill and outputs.

This scenario challenges the very foundations of the current understanding of creativity. As the boundaries of collaboration between human and AI

DOI: 10.4324/9781003453901-5

systems blur during co-creative processes, there is a need for a profound re-evaluation of current assumptions, definitions, and frameworks of creativity. The impersonal and autonomous nature of AI introduces a novel dynamic to creative processes, demanding a departure from conventional paradigms, which is heavily based on sole human elements. Thus, it could not be more different from the, often, emotionally charged, subjective nature of human creativity (especially in artistic sectors). Algorithms lack personal biases and emotional influences, as they generate creative outputs based on patterns, previously existing data, and learning processes.

Thus, the autonomy of AI systems raises fundamental questions about the nature of creativity itself. For example, how to attribute intentionality to AI-generated outputs when there is an absence of human consciousness, intentionality, and subjective experience involved in the process. Therefore, the existing creativity frameworks for understanding creativity must evolve to incorporate the nuances introduced by AI, discussing the connection between human and AI in co-creative processes, recognizing the symbiotic relationship that unfolds in such cases.

This chapter proposes a critical reflection and the need to adapt (or extend) current assumptions found in existing creativity frameworks, such as the 8P creativity framework proposed by Sternberg and Karami (2022), to account for the impersonal and autonomous nature of autonomous systems.

A Brief Overview of Traditional Creativity Frameworks

Creativity is often described as the skill of creating ideas or artefacts that are new, valuable, and surprising (Boden, 2004; Kaufman & Sternberg, 2019). As it represents a complex concept, authors have long developed frameworks, in the attempt to create a holistic overview and help elucidate theories that explain it. Some of the most prominent creativity frameworks include the 4P framework by Rhodes (1961); the 5A framework by Glăveanu (2013); the 7C framework by Lubart and Thornhill-Miller (2019); and, most recently, the 8P framework by Sternberg and Karami (2022).

The framework developed by Rhodes's (1961) represents a seminal work, on which many of the frameworks that followed are grounded. It is widely known as the 4P framework and consists of the following elements:

- **Persons:** *Characteristics of creative individuals such as personality, intellect, temperament, and traits.*
- **Process:** *The mental operations that produce creative work. It involves, for example, motivation, perception, learning, thinking, and communication.*
- **Press:** *Environmental forces that incite people to produce creative work.*
- **Product:** *The outcome of creativity, such as an idea, process, or physical object.*

Following the work of Rhodes (1961), authors have later proposed extensions to the 4Ps. In other words, they have adopted the premises of the 4Ps (often suggesting reconsiderations) and introduced new elements. For example, the 5A framework by Glăveanu (2013) extended the element of *press* by "dividing" it into *affordance* (relationship between the creator and the physical environment) and *audience* (social environment). Later, Lubart and Thornhill-Miller (2019) proposed the 7C framework, incorporating *consumption* (adoption of creative products, either into daily life or by endorsing them), *collaboration* (processes engaged when individuals collaborate on creative initiatives), and *curriculum* (forms of teaching creativity to develop it in people). And most recently, Sternberg and Karami (2022) proposed the 8P framework, where the authors discussed *propulsion* (the way in which, and the rate at which, creativity changes people's thinking), which had been previously suggested by the author (Sternberg, 1999), and introduced two new additional factors: *propulsion* (the way in which, and the rate at which, creativity changes people's thinking) and *public* (the audience of creative work).

Table 3.1 depicts an overview of how different frameworks overlap and how each proposed to extend Rhodes's (1961) 4P framework.

The 5A, 7C, and 8P frameworks were not the only attempts to extend the initial model from Rhodes (1961); however, including all additional attempts goes beyond the scope of this chapter. Nevertheless, some examples include Simonton (1995), who proposed adding the element of *persuasion* (the influence of creativity with a sense of leadership), and Runco (2007), who proposed applying a hierarchical structure to the framework and classifying the elements into the *creative potential* (containing *person, process*, and *press*) and the creative performance (containing *product* and *persuasion*).

The premises of all elements discussed in the existing "traditional" frameworks for creativity have been instrumental to the understanding of concepts and theories related to it across various fields, which highlights their unquestionable relevance. Nevertheless, in face of the revolution caused by AI in creative and non-creative sectors, they face an important challenge: their premises were developed focusing on humans as the main and sole agent of creative processes and deeply rooted human sentiments for both creation and judgement or perception of the creative output. For example, the assumptions for *persons* (*actor or creator*) are centred on human physical and mental abilities. *Process* is also often discussed as "mental" operations or processes, also from a human perspective. The assumptions for *person* and *process*, as creative elements, are directly linked to the understanding of skill (Amabile, 1996) and expertise (Baer, 2015), both required for creative processes and vital for the value perception (by the *public* or *audience*) of a creative output (*product*, *artefact*, or *creations*).

Therefore, as the characteristics and motivations of the creative agent changes, from *human* (organic and personal) to a *co-creative process between a human and AI* (partially organic and personal and partially autonomous

Table 3.1 Overview of the 4P (Rhodes, 1961), 5A (Glăveanu, 2013), 7C (Lubart & Thornhill-Miller, 2019) and 8P (Sternberg & Karami, 2022) creativity framework elements.

Framework Elements				*Description*
4Ps	*5As*	*7Cs*	*8Ps*	
Persons	Actor	Creators	Persons	Characteristics of creative individuals, such as personality, intellect, temperament, and trait.
Process	Action	Creations	Process	The mental operations that produce creative work.
Press	Audience	Context	Press	Environmental forces that incite people to produce creative work.
Product	Artefact	Creations	Product	The outcome of creativity, e.g. idea, process, or physical object.
Additions to Rhodes's 4Ps	Affordance	–	–	Influences from the physical environment towards creative expressions.
		Consumption	–	Adoption of creative products, either into daily life or by endorsing them.
		Collaboration	–	Processes engaged when individuals collaborate on creative initiatives.
		Curriculum	–	Forms of teaching creativity to develop it in people.
			Public	The actor's engagements with others, encompassing external factors that motivate the creator's efforts.
			Propulsion	The way in which, and the rate at which, creativity changes people's thinking.
			Purpose	What creativity is for?
			Problem	The tasks and situations requiring creative solution.

and impersonal), or *autonomously by an AI system* (autonomous and impersonal), many of the existing conceptual premises described in "traditional" creative frameworks no longer stand. Consequently, there is a critical need to adapt them to account for the autonomous and impersonal nature of AI and the nature of different forms of artificial co-creative processes. Nevertheless, the field of computational creativity has contributed to this discussion by developing frameworks grounded on premises of the 4Ps, attempting to account for the interlink between humans and AI, during artificial co-creative processes. Some of these models are discussed next.

Creativity Elements and AI

Jordanous (2016) provided one of the initial contributions to the adaptation of creativity framework models to computational creativity, by adjusting to premises from Rhode's (1961) 4Ps model. The author defended that "broader views of creativity afforded by the Four Ps is vital in guiding us towards more comprehensively useful computational investigations of creativity" (p. 194). Thus, Jordanous (2016) considered each element to make evaluations of novelty and value (core pillars of the definition of creativity, according to Boden, 2004 and many others). This proposal was then later adopted in further works, such as Corneli (2016) and Kantosalo (2019).

In short, the main conceptual suggestions proposed by the author focused on two elements: *person* and *process*. For example, Jordanous (2016) considered that, for computational creativity, *person* should be viewed as *producer*, as it could be a "computer program, software, robot or a creative agent within a multi-agent system" (p. 197). She also advocated for broadening the understanding of how the *process* should understood, as the "creative process might be the algorithms employed by a single piece of software, or interactions between multiple machines or programs, or interactions between machine and human users, or between a machine and the environment it is embodied in" (p. 198).

However, it was Kantosalo and Takala (2020) who proposed a fully novel framework for human–computer co-creativity, labelled as the "5C". It was grounded on the traditional human-centred frameworks of Rhode's (1961) and Glăveanu's (2013), and on the proposed adaptation of Jordanous (2016) to computational creativity. After an extensive review, Kantosalo and Takala (2020) proposed the following five elements:

- **Collective**: *Entities actively involved in the co-creative collaboration – usually, one human and one computational collaborator.*
- **Collaboration**: *Creative processes of the collaborators and how these are fitted together to form a collective creative process.*
- **Contributions**: *Complete and incomplete, tangible and intangible, creative artefacts which are exchanged during the co-creative process.*

- **Community:** *Social environment, which affects the design of co-creative systems.*
- **Context:** *Material surroundings of the work, which affect the design of co-creative systems.*

Moreover, the literature on human–AI co-creation (or collaboration) frameworks also presents various insights which could well be incorporated into the understanding of artificial creativity. For example, building on decades of research on human–AI collaboration, Muller and Weisz (2022) proposed two novel and extremely relevant elements: *dynamism* and *sociality*. The authors detail that human–AI collaborations present *dynamic* changes and interactions between human and smart systems, which are pivotal for the output which is generated. Also, that such dynamism is dependent on the type and level of co-creation. After all, as Parasuraman et al. (2000) details, the levels of human–AI collaboration should be viewed as a spectrum ranging from complete computer autonomy to complete human autonomy. And that between these two extremes, there are varying levels and forms of possible human–AI collaboration, where each one may have different levels of influence on the creative process (named "the human-computer trade-off allocation"). And that these interactions may occur in a flexible and dynamic form, with possible constant changes in initiative and autonomy (Muller & Weisz, 2022). Such *dynamism* during the creative process is much aligned with Kantosalo and Takala's (2020) and Lubart and Thornhill-Miller's (2019) elements of *collaboration*.

Reflecting on Missing Elements from Traditional Creativity Frameworks for AI

Despite the comprehensiveness of such models, a series of elements from traditional creativity models still have not been adjusted to the characteristics of AI. For example, *propulsion*, *purpose*, and *problem* (as proposed by Sternberg & Karami, 2022), reveal an important gap in the current literature. Thus, some considerations are presented next.

Propulsion

Propulsion, as described by Sternberg and Karami (2022), relates to "the way in which, and the rate at which creativity changes people's thinking" (p. 58) and may be accomplished in three main types: (1) *paradigm-preserving creators*, where creators work creatively within an existing paradigm or set of paradigms; (2) *paradigm busters*, where creators break away from conventional norms; and (3) *paradigm synthesizers*, where creators combine different existing paradigms to come up with a novel contribution.

To achieve the different levels of propulsion, the mode of the creative process is crucial. Boden (2004, 2007), for example, described three main modes of creativity which help understand the different forms of creative processes: *combinatory* creativity (unfamiliar combination of familiar ideas), *exploratory* creativity (generating new outcomes within existing boundaries or rules), and *transformational* creativity (breaking away from existing boundaries or rules to generate novel outcomes). These different modes enable the development of outputs with varying levels of novelty: *historical* novelty (when something is novel to a society at large) and *psychological* novelty (novel to a single individual, but not necessarily to others).

Mastering these three modes of creativity is vital to generate outputs with varying forms of *propulsion*, as described by Sternberg and Karami (2022). In view of the extraordinary computing capacity of AI and consequently the possibility of generating outputs at a scale which is unachievable for humans alone, all three modes of creativity (*combinatory, exploratory, and transformational*) can be achieved at a much larger rate than attempted by sole human creators (Gruner & Csikszentmihalyi, 2019). Consequently, AI-generated, or co-created, outputs are much more likely to impact both psychological novelty (which is easier to achieve) and historical novelty (much rarer), thus generating greater *propulsion* in society (e.g. a new music genre, which can impact and shape new cultural symbols and values), as described by Boden (2004) and Picciuto and Carruthers (2014).

Problem

Other element that requires a critical reflection is *problem*, which, as suggested by Sternberg and Karami (2022), refers to tasks and situations requiring creative solution. The authors explained that creativity is domain-specific, and that it becomes even more domain-specific the more it moves from *mini*-C ("novel and personally meaningful interpretation of experiences, actions, and events", Beghetto & Kaufman, 2007, p. 73) to *little*-C (everyday innovation), then to *Pro*-C (creative at a professional level) and finally to *Big*-C (domain-changing innovations and accomplishments), as described by Kaufman and Beghetto (2009). Moreover, the authors claimed that humans tend to be creative in specific domains rather than in multiple domains, as one needs great in-depth knowledge of a domain to identify a problem and innovative ways of addressing them (Weisberg, 2006). Consequently, few individuals are creative at a *Pro*-C level, and it is extremely rare for some to be at a *Big*-C level (e.g. Darwin, Picasso, Einstein, and Beethoven).

AI has shown to act extremely efficiently in identifying problems and suggesting creative solutions at all creative levels and at various domains (extremely unlikely for humans, as high-level skill and expertise are required). Examples include flood forecasting (Adikari et al., 2021), real-time wildfire detection (Wahyono et al., 2022), monitoring perinatal health (Ramakrishnan et al., 2021), pest monitoring (Batz et al., 2023), genetic variations that may cause diseases (Ahmed et al., 2021), and more.

Furthermore, authors have predicted clear evolutionary development stages for AI (Antonov, 2011). Currently, most AI systems act on what is known as "narrow AI" (Todorova, 2020), which means achieving a super-human performance on a single task (e.g. write text or playing chess). The next stage is "general" AI (although some authors defend this might never happen, such as Fjelland, 2020), when the system has super-human-like capabilities across different domains. The final predicted stage is called "super" AI (Gill, 2016), when it outperforms human intelligence across all domains simultaneously. This suggests that one of the main innovations in the development of AI is that it will perform across all possible domains and be able to master the identification of problems and solutions in a creative way (novel and valuable) at *Big*-C level, simultaneously. For example, as a single entity capable of being an extraordinary music composer, airplane pilot, cook, writer, designer, doctor, engineer, statistician, chemist, philosopher, and much more. All at once.

Purpose

As previously discussed, creativity is the act of developing outputs which are regarded as novel, surprising, and valuable (Boden, 2004). But the act of being creative might serve different purposes. For example, the *purpose* of creativity is often discussed in terms of its direction: *positive*, *negative* (also referred to as "dark side of creativity"), and *mixed* (Clark & James, 1999; Cropley et al., 2010, 2014). A purpose is deemed as *positive*, when applied in a constructive manner or beneficial for a domain or society at large (Sternberg, 2021; Sternberg & Karami, 2022). Contrarily, it is considered *negative* when serving a purpose which is destructive for a domain or society. And it is defined as *mixed*, when the purpose may be viewed as either positive or negative, depending on the domain or context in which it is applied.

Traditionally, creative purposes are defined by motivations which are intrinsic to humans. For example, Sternberg and Karami (2022) describe that individuals might engage in *positive* creativity not just to improve their reputation to others but often, and significantly, to contribute positively, meaningfully, and potentially enduringly to a specific domain or others. Furthermore, motivations for *negative* creativity might include: enhancing self-reputation by damaging others, influence of strong emotions (anger, fear, sense of revenge), or even for the recognition of other for being evil.

However, algorithms are not human entities. They are created by humans, but often act independently according to its own intelligence. Terzidis et al. (2023), for example, considers that the context is relevant to define intentionality. The authors make a distinction between contexts where there is an external goal (normally defined by humans, such as how a self-driving car should make decisions) and contexts where there is no external goal (such as art, as it is a form of free expression). The authors defend that, in the case of arts, what is most relevant is the "intentio" found in the creative product, rather than the process through which it was made. Consequently, intentionality would not require a human influence and could therefore be "unintentional". They call this "unintentional intentionality". This contradicts, however, the premises defended by Sternberg and Karami (2022) for *positive*, *negative*, and *mixed* purposes, which are fundamentally grounded on human motivation and emotions.

Remaining Elements

Finally, a few remaining elements such as *consumption* and *curriculum* (proposed by Lubart & Thornhill-Miller, 2019) and *affordance* (Glăveanu, 2013) have not been discussed in this section, but it also require future studies to critically reflect on their nature considering the characteristics of AI.

Furthermore, as the process of human–AI collaboration evolves and matures due to both technological advancements and wide adoption, new elements are expected to arise that will help explain the novel forms of human–AI co-creation and their impact. It is therefore essential to permanently investigate the nature of such processes. This chapter will later address some of these issues in the last section while proposing a future research agenda.

Future of Creativity Frameworks

The main assumptions of traditional creativity frameworks hold a paradigm which does not account for the autonomous and impersonal nature autonomous systems nor the intricacies of artificial co-creative processes. They are grounded on principles of a pre-AI revolution era, where the ability to be creative was deemed to be restricted to humans. On the other hand, the perspectives defended by Jordanous (2016), Kantosalo and Takala (2020), and Muller and Weisz (2022), to name a few, expand on these assumptions, to account for a more current understanding of creativity. With the advances of generative AI tools, capable of creating high-quality creative outputs such as writing, music composition, painting, filmmaking and animations, 3D artworks, and more, the adoption rate is only expected to increase. Thus, we should also expect innovative forms of co-creation within the "the human-computer trade-off allocation" as proposed by Parasuraman et al. (2000). Consequently, in view of such fast adoption, it will be impossible to distinguish if a creative output has been solely produced by humans, autonomously by AI or the co-creative collaborative forms in which it was made.

Therefore, it seems unreasonable to maintain premises for the understanding of creativity that disregard this new reality and that are restricted to a world where humans are the main (and possibly only) creative agent. Instead, the understanding of creativity should adopt a new paradigm (Tigre Moura, 2023), one where smart systems are seen as creative agents and one where all the characteristics of artificial creativity are included.

Future Research Agenda

The discussion presented in this chapter not only delves into the intricacies of frameworks of creativity and AI but also opens the door to a series of future research questions. These questions focus on the intersection of creativity and AI and hopes to tackle existing gaps in the current literature. Addressing them will hopefully provide insights into the dynamic relationship between human creativity and the evolving landscape of AI. Next, I present three questions for future research:

RQ1: To what extent should the field of creativity adopt computational creativity premises as a new paradigm, and how can existing frameworks be adapted to embrace smart systems as agents in the creative process?

Creative processes and agents have undoubtedly changed with the advent and wide adoption of smart systems. Thus, the adoption of premises from computational creativity models for the understanding of creativity and extending previously established frameworks will allow for a truly holistic overview of the topic. This will enable the understanding of concepts and theories of creativity in the age of artificial co-creation and as a "new normal" for creative processes. Moreover, future research should focus on conducting exploratory investigations to identify potential novel elements that are not found in current traditional creativity frameworks nor in computational creativity ones.

RQ2: How will varying levels of human–AI co-creation during creative processes impact the value perception of created outputs?

Value represents a fundamental factor in most definitions of creativity (Boden, 2004), although some authors might debate it (Weisberg, 2015; Harrington, 2018). The value perception of creative outputs is highly subjective and context based, for example, by product type (Tigre Moura et al., 2023). Importantly, it is set by factors determined by the *audience* (Glăveanu, 2013) and or *public* (Sternberg & Karami, 2022), such as skill, expertise, and intentionality. However, as the adoption of smart systems may impact such factors (e.g. lower importance of creative skill, such as playing an instrument for music composition), it is crucial to understand how this will affect the perceived value of outcomes. Previous studies, such as in music (Tigre Moura & Maw, 2021), have shown that AI-generated, or co-created, outputs can rival human-made ones. However, wide adoption will also lower a positive perception which may be associated with an innovative process (co-creation with AI). Future research should therefore investigate if, for example, value will be defined solely by the response it causes on humans (e.g. emotional response of a song or painting), rather than by how the creative process was conducted and the role of humans in it.

As such artificial co-creation practices become widely adopted, and there is an abundance of creative outputs, this new reality is expected to impact the value of outputs. The greater the offer, potentially the lower will be the value. Thus, future research should also focus on developing a longitudinal understanding of the value of AI-generated, or co-created, outputs, across markets, cultures, contexts (e.g. high- or low-involvement products).

RQ3: To what extent does the involvement of AI in creative processes impact the self-identity of human creatives, and how does this influence the authenticity and uniqueness attributed to their creative outputs?

As AI becomes an integral part of the creative journey, human creatives are faced with a transformative shift in their creative roles, one where the creative skill and expertise, for example, may no longer be necessary. With such democratization of creative processes, non-creatives will also be able to generate outputs which will be perceived as novel and valuable by the public. Consequently, such change is expected to impact the self-identity of creatives, as ones with a unique ability and expertise which is admired and respected in society (e.g. to master a painting technique or musical instrument). Addressing this question will focus on psychological and existential issues of creatives, examining how human creators may reconcile their identity and artistic agency when collaborating with AI.

References

Adikari, K. E., Shrestha, S., Ratnayake, D. T., Budhathoki, A., Mohanasundaram, S., & Dailey, M. N. (2021). Evaluation of artificial intelligence models for flood and drought forecasting in arid and tropical regions. *Environmental Modelling & Software*, *144*, 105136.

Ahmed, H., Alarabi, L., El-Sappagh, S., Soliman, H., & Elmogy, M. (2021). Genetic variations analysis for complex brain disease diagnosis using machine learning techniques: Opportunities and hurdles. *PeerJ Computer Science*, 7, e697.

Amabile, T. M. (1996). *Creativity and innovation in organizations* (Vol. 5). Harvard Business School.

Antonov, A. A. (2011). From artificial intelligence to human super-intelligence [J]. *Artificial Intelligence*, 2(6), 3560.

Baer, J. (2015). The importance of domain-specific expertise in creativity. *Roeper Review*, *37*(3), 165–178.

Batz, P., Will, T., Thiel, S., Ziesche, T. M., & Joachim, C. (2023). From identification to forecasting: The potential of image recognition and artificial intelligence for aphid pest monitoring. *Frontiers in Plant Science*, *14*, 1150748.

Beghetto, R. A., & Kaufman, J. C. (2007). Toward a broader conception of creativity: A case for "mini-c" creativity. *Psychology of Aesthetics, Creativity, and the Arts*, *1*(2), 73.

Boden, M. A. (2004). *The creative mind: Myths and mechanisms*. Routledge.

Boden, M. A. (2007). Creativity in a nutshell. *Think*, *5*(15), 83–96.
Clark, K., & James, K. (1999). Justice and positive and negative creativity. *Creativity Research Journal*, *12*(4), 311–320.
Corneli, J. (2016). An institutional approach to computational social creativity. In *Proceedings of the seventh international conference on computational creativity* (pp. 131–138). UPMC.
Cropley, D. H., Cropley, A. J., Kaufman, J. C., & Runco, M. A. (Eds.). (2010). *The dark side of creativity*. Cambridge University Press.
Cropley, D. H., Kaufman, J. C., White, A. E., & Chiera, B. A. (2014). Layperson perceptions of malevolent creativity: The good, the bad, and the ambiguous. *Psychology of Aesthetics, Creativity, and the Arts*, *8*(4), 400.
Fjelland, R. (2020). Why general artificial intelligence will not be realized. *Humanities and Social Sciences Communications*, *7*(1), 1–9.
Gill, K. S. (2016). Artificial super intelligence: Beyond rhetoric. *AI & Society*, *31*, 137–143.
Glăveanu, V. P. (2013). Rewriting the language of creativity: The five A's framework. *Review of General Psychology*, *17*(1), 69–81.
Gruner, D. T., & Csikszentmihalyi, M. (2019). Engineering creativity in an age of artificial intelligence. In *The Palgrave handbook of social creativity research* (pp. 447–462). Palgrave.
Harrington, D. M. (2018). On the usefulness of "value" in the definition of creativity: A commentary. *Creativity Research Journal*, *30*(1), 118–121.
Jordanous, A. (2016). Four pppperspectives on computational creativity in theory and in practice. *Connection Science*, *28*(2), 194–216.
Kantosalo, A. (2019). *Human-computer co-creativity: Designing, evaluating and modelling computational collaborators for poetry writing* [PhD dissertation, University of Helsinki], Report A-2019-3.
Kantosalo, A., & Takala, T. (2020, September). *Five C's for human-computer co-creativity-an update on classical creativity perspectives* (pp. 17–24). ICCC.
Kaufman, J. C., & Beghetto, R. A. (2009). Beyond big and little: The four c model of creativity. *Review of General Psychology*, *13*(1), 1–12.
Kaufman, J. C., & Sternberg, R. J. (Eds.). (2019). *Cambridge handbook of creativity* (2nd ed.). Cambridge University Press. https://doi.org/10.1017/9781316979839
Lubart, T., & Thornhill-Miller, B. (2019). Creativity: An overview of the 7C's of creative thought. In *The psychology of human thought: An introduction* (pp. 277–306). Heidelberg University Publishing (heiUP).
Muller, M., & Weisz, J. (2022, June). Extending a human-AI collaboration framework with dynamism and sociality. In *2022 symposium on human-computer interaction for work* (pp. 1–12). ACM.
Parasuraman, R., Sheridan, T. B., & Wickens, C. D. (2000). A model for types and levels of human interaction with automation. *IEEE Transactions on Systems, Man, and Cybernetics-Part A: Systems and Humans*, *30*(3), 286–297.
Picciuto, E., & Carruthers, P. (2014). The origins of creativity. *The Philosophy of Creativity: New Essays, 199.*
Ramakrishnan, R., Rao, S., & He, J. R. (2021). Perinatal health predictors using artificial intelligence: A review. *Women's Health*, *17*. https://doi.org/10.1177/17455065211046132
Rhodes, M. (1961). An analysis of creativity. *The Phi Delta Kappan*, *42*, 305–310.
Runco, M. A. (2007). A hierarchical framework for the study of creativity. *New Horizons in Education*, *55*(3), 1–9.
Simonton, D. K. (1995). Exceptional personal influence: An integrative paradigm. *Creativity Research Journal*, *8*(4), 371–376.

Sternberg, R. J. (1999). A propulsion model of types of creative contributions. *Review of General Psychology*, *3*(2), 83–100.
Sternberg, R. J. (2021). Positive creativity. In A. Kostic & D. Chadee (Eds.), *Current research in positive psychology* (pp. 33–42). Palgrave Macmillan.
Sternberg, R. J., & Karami, S. (2022). An 8P theoretical framework for understanding creativity and theories of creativity. *The Journal of Creative Behavior*, *56*(1), 55–78.
Terzidis, K., Fabrocini, F., & Lee, H. (2023). Unintentional intentionality: Art and design in the age of artificial intelligence. *AI & Society*, *38*(4), 1715–1724.
Tigre Moura, F. (2023). Artificial intelligence, creativity, and intentionality: The need for a paradigm shift. *The Journal of Creative Behavior*, *57*, 336–338. https://doi.org/10.1002/jocb.585
Tigre Moura, F., Castrucci, C., & Hindley, C. (2023). Artificial intelligence creates art? An experimental investigation of value and creativity perceptions. *The Journal of Creative Behavior*, *57*, 534–549. https://doi.org/10.1002/jocb.600
Tigre Moura, F., & Maw, C. (2021). Artificial intelligence became Beethoven: How do listeners and music professionals perceive artificially composed music? *Journal of Consumer Marketing*, *38*(2), 137–146.
Todorova, M. (2020). "Narrow AI" in the context of AI implementation, transformation and the end of some jobs. *Nauchni trudove*, *4*, 15–25.
Wahyono, Harjoko, A., Dharmawan, A., Adhinata, F. D., Kosala, G., & Jo, K. H. (2022). Real-time forest fire detection framework based on artificial intelligence using color probability model and motion feature analysis. *Fire*, *5*(1), 23.
Weisberg, R. W. (2015). On the usefulness of "value" in the definition of creativity. *Creativity Research Journal*, 27(2), 111–124.
Weisberg, R. W. (2006). *Creativity: Understanding innovation in problem solving, science, invention, and the arts*. Wiley.

Chapter 4

Creative Sense-Making

A Cognitive Framework for Modelling Interaction Dynamics in Co-Creative AI

Nicholas Davis

Introduction

The field of co-creative artificial intelligence (AI) designs, implements, and develops evaluation methods for co-creative agents. A co-creative agent is an agent that collaborates with one or more users on a shared creative product (Davis, 2013). The contributions can be mutually influential (Davis, 2013), share creative responsibility in the co-creative process (Kantosalo et al., 2014; Kantosalo & Takala, 2020), and have a mixed initiative between the user and the system (Yannakakis et al., 2014). Collaborators engage in open-ended meaning construction that emerges through interaction (Davis, 2013). Instead of replacing human creativity, co-creation engenders collaboration between humans and AI. There are examples of co-creative systems in several creative domains, such as dance, design (Karimi et al., 2019; Ibarrola et al., 2022), drawing (Davis et al., 2015), game design (Guzdial et al., 2018), music (Biles, 2002), poetry (Kantosalo & Riihiaho, 2019), and theatrical improvisation (Magerko et al., 2011).

The creative sense-making (CSM) framework models the interaction dynamics of co-creation in the context of co-creative AI (Davis et al., 2017). It proposed three basic cognitive interaction modes a user can engage in: *explore* (termed physical unclamp), *execute* (termed clamped), and *wait* (termed mental unclamp). Davis et al. (2017) used a video coding tool that enables the researcher to apply codes continuously through time through a selection mechanism, such as a slider or hot keys that change the interaction mode code. The video coding tool samples from the user's code application slider at specified time intervals during video playback (e.g. every 0.5 seconds) to produce a continuous stream of data. The procedure codes modes of interaction continuously through time to generate a sense-making curve that illustrates the interaction dynamics of the co-creative process.

Recently, the CSM was extended by Deshpande et al. (2023) as observable creative sense-making (OCSM). A new coding convention was proposed that uses *newness*, *participation*, and *appropriateness* to evaluate interactions continuously. Each category has four states with behavioural markers

DOI: 10.4324/9781003453901-6

for each state, producing three distinct sets of data. These datasets are analysed to compare the interactions between dancers in a dyadic improvisational dance context. The new coding approach captures dimensions of co-creation not covered by the initial CSM framework. However, the authors entirely dispense with the CSM's coding technique, when the two could be used in tandem. The CSM captures raw interaction data that are continuous, whereas the OCSM captures more qualitative data that are one level above the raw interaction data. This chapter explains how the CSM and OCSM curves can be used in tandem to provide a more holistic view of the interaction dynamics of co-creation.

This chapter reviews the enactive cognitive science theory of participatory sense-making (PSM) upon which the CSM and OCSM are based. Then, the CSM is described, including explaining its theoretical foundations, describing its coding scheme, and demonstrating the data it generates. Next, the chapter describes the OCSM and explains its coding conventions. The authors of the OCSM reported difficulty in applying the CSM coding convention. Thus, these difficulties are addressed in the section about the CSM. After this, the chapter describes the extended CSM framework that includes the new features added by OCSM: newness, appropriateness, and participation. Then, the CSM coding scheme is developed for several domains of co-creative AI. A discussion follows that examines how the CSM can be used to automatically code interactions in co-creative AI systems leading to a new type of *quantified co-creative AI system* that models the interaction dynamics of co-creation happening on the platform.

The main contributions of this chapter are extending the CSM framework to include recent work in the field. The chapter also simplifies and clarifies the CSM coding procedure, which should facilitate its adoption. A high-level coding scheme for multiple co-creative AI domains is provided to aid researchers in modelling co-creation using the CSM.

Enactive Sense-Making Frameworks

Participatory Sense-Making

The cognitive science theory of enaction proposes a radical new conceptualization of how humans build and construct meaning in a process called *sense-making*. Sense-making is the process of a self-organizing system regulating its interaction with the environment by casting a web of significance through interaction with that environment (De Jaegher & Di Paolo, 2007). Importantly, during sense-making, meaning emerges through embodied (Varela et al., 2017) and situated interaction with the environment. Instead of high-level goals and scripts determining action, agents engage in sense-making to dynamically construct meaning through the coordination of interactions and the feedback from those interactions.

When multiple people are involved in this sense-making process, it is referred to as *PSM*, which is defined as "*the coordination of intentional activity in interaction, whereby individual sense-making processes are affected, and new domains of social sense-making can be generated that were not available to each individual on their own*" (De Jaegher & Di Paolo, 2007 p. 497). This coordination can be directed towards regulating the interaction, referred to as functional coordination, or contributing to the interaction, referred to as interactional coordination (De Jaegher & Di Paolo, 2007). Kellas and Trees (2014) apply this distinction to co-creation to arrive at a model of interactional sense-making that has two distinct sense-making processes: interactions between collaborators and interactions with the shared product.

When collaborators are co-regulating both interactional sense-making processes, PSM emerges (Rezwana & Maher, 2022). During co-creation, participants actively engage in PSM as they build meaning and negotiate shared intentions in the creative product (Davis, 2013). During PSM, interactions are coupled such that each subsequent interaction is influenced by what came before (De Jaegher & Di Paolo, 2007). This coupling has a depth (the number of turns), a duration (total time spent coupled), an initiator (the individual who began the coupling), and a decoupler (the individual who ended the coupling).

CSM Framework

The CSM framework quantifies and models the interaction dynamics of co-creation using the cognitive science theory of enaction. This framework was initially presented by Davis et al. (2017). In this chapter, the CSM is refined to explicitly include an AI in the interaction. The CSM framework proposes cognitive modes that fluctuate through time to form sense-making processes in the creative process. There are two main cognitive modes that form the basis of the CSM framework: *clamped* and *unclamped* cognition. Clamped cognition is defined as a cognitive agent having a well-developed mental model of the situation, knowing what to do, and fluidly executing on-task actions. It corresponds to the state of creative flow identified in the psychological literature on creativity (Csikszentmihalyi et al., 2014). When something unexpected happens, a problem arises, or the cognitive agent wants to reframe the creative context, cognition is said to unclamp, and the agent builds a new mental model of the situation through exploratory actions and feedback from the environment and other agents within it. There are two types of unclamped cognition: *functional unclamp* and *interactional unclamp*. Functional unclamp refers to regulating the interaction with the environment and other agents within it. There is a partial functional unclamp, which is manipulating the interface of the tool to gather more resources for the creative activity (e.g. changing brush stroke colour), and there is a full functional unclamp, which is communicating to a collaborator (e.g. requesting inspiration from the AI agent). Interactional unclamp refers to disengaging from the interaction. These three cognitive modes lead to four modes of interaction that are coded in the CSM framework: *execute* (clamped cognition), *interface manipulation* (partial functional unclamp), *communicate* (functional unclamp), and *wait* (interactional unclamp).

There are two main interactional processes occurring in co-creative AI: interaction between collaborators (functional coordination) and interaction with the shared product (interactional coordination) (Kellas & Trees, 2014; Rezwana & Maher, 2022). Figure 4.1 shows functional coordination is split into two codes: communicating with the agent and gathering resources for the creative activity, such as interacting with the interface of a system, for example changing brush stroke colour. There are two types of communication channels in co-creative AI, as shown in Figure 4.1: human-to-AI communication and AI-to-human communication (Rezwana & Maher, 2022). Each communication channel can be used to convey different types of information, such as instructions, requests, explanations, feedback, and ideations. Interactional coordination is split into two phases: *execute* and *wait*. These cycles of execution and waiting form emergent turn-taking. An interaction rhythm emerges that coordinates interaction in PSM (De Jaegher & Di Paolo, 2007). By coding both execution and waiting periods, it is possible to begin to model this interaction rhythm and the emergent turn-taking that results. The three cognitive modes of the CSM are based on a theory and model of cognition called the enactive model of creativity (Davis et al., 2015).

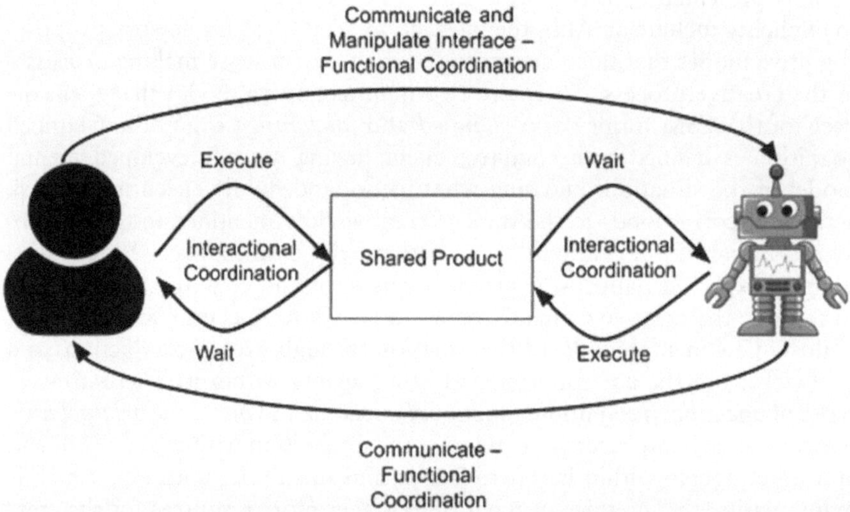

Figure 4.1 CSM coding convention mapped to the Kellas and Trees (2014) interactional sense-making with coordination processes from De Jaegher and Di Paolo (2007).

In the enactive model of creativity, the awareness of the agent is a dynamic component that shifts along the horizontal axis during cognition. When awareness is centred, it is at equilibrium and cognition is said to be clamped. Here, actions are effortlessly executed. This is possible due to having a well-developed *perceptual logic* (Davis et al., 2015), which is a mechanism for selecting relevant object affordances (Norman, 2013; Gibson, 2014) from all potential affordances (Norman, 1999). In clamped cognition, one can "read" the environment in the sense that affordances in the environment guide interactions without the need for a complex mental model of the situation. When something surprising happens, the current strategy breaks down, or the agent wants to reframe the interaction in some manner, the agent has to unclamp cognition, which would be depicted as the awareness rectangle moving either right or left. During unclamped cognition, the individual can work to build an explicit mental model during cycles of exploration and feedback from the environment, including communicating with other cognitive agents in the environment. These clamping and unclamping events form the basis for the CSM interaction modes. In the context of co-creative AI, the coding scheme that arises is: clamped cognition is *executing* fluid actions, unclamped in the mental domain is *waiting*, and unclamped in the environmental domain is *communicating* and *interface manipulation*.

In the CSM, each cognitive interaction mode has behavioural markers that can be used to detect its presence during the coding process. Each interaction mode is given a value during the coding process. The interaction mode of the agent is continuously coded through time to provide the raw interaction data from which we can determine different types of sense-making. With these data, it is possible to compare the sense-making activities between participants of a session and between sessions.

Figure 4.2 shows the raw interaction data from a five-minute co-creative drawing session. The number of code applications can be counted to quantify

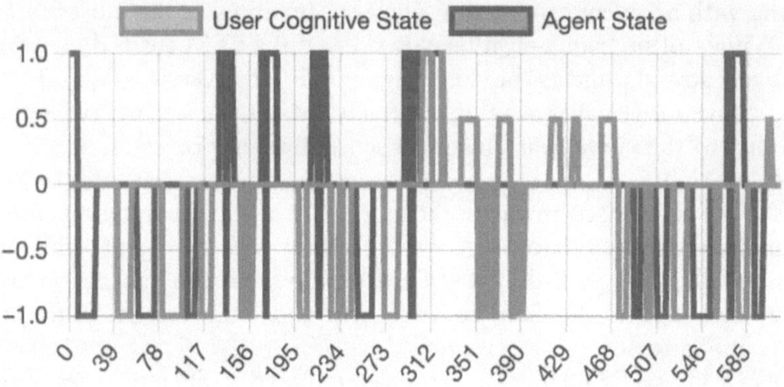

Figure 4.2 Raw interaction data from a five-minute co-creative drawing session.
Note: *X*-axis: time; *Y*-axis: code applied.

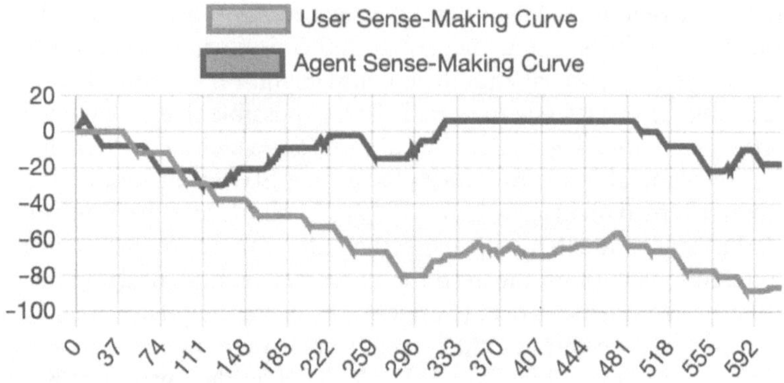

Figure 4.3 The CSM curve from a five-minute co-creative drawing session.

Note: *X*-axis: time; *Y*-axis: cumulative sum of interaction codes.

the session. These data can also be summed to further identify trends, as shown in Figure 4.3.

In Figure 4.3, different trends can be identified, such as oscillating cycles, upward trends, horizontal trends, and downward trends. The novelty and value of the CSM is in the continuous nature of the data it provides about interaction dynamics. With these continuous data, it is possible to do statistical analysis and machine learning to identify trends and patterns in the data.

The previous CSM coding scheme was as follows: clamped is 0, functional unclamp is 1, and interactional unclamp is –1. The CSM codes a partial functional unclamp as 0.5. The updated CSM coding scheme is as follows: clamped is –1, functional unclamp is 1, partial functional unclamp is 0.5, and interactional unclamp is 0. The reason for this change in code application procedure is because Davis (2023) found that in the domain of co-creative drawing with an agent, interaction unclamp (waiting) is the most dominant code. With waiting being –1, it heavily skewed the CSM curve downward in a relatively straight line as shown in Figure 4.4 (left) (Davis, 2023). The new coding convention makes waiting 0, which doesn't skew the curve and adds variability to the curve that illustrates sense-making processes, as shown in Figure 4.4 (right).

When the curve rises upwards, the user is either manipulating the interface or regulating the interaction by communicating with the agent, while when the curve falls, the user is drawing. Events from the user's interaction history can be overlaid onto the CSM curve to help understand trends in the curve, such as in Figure 4.5.

The raw interaction data can be utilized to identify different types of sense-making in the co-creative process, as shown in Figure 4.6. Exploratory sense-making is characterized by multiple sense-making cycles of functional unclamp and clamped execution. Functional coordination has several functional

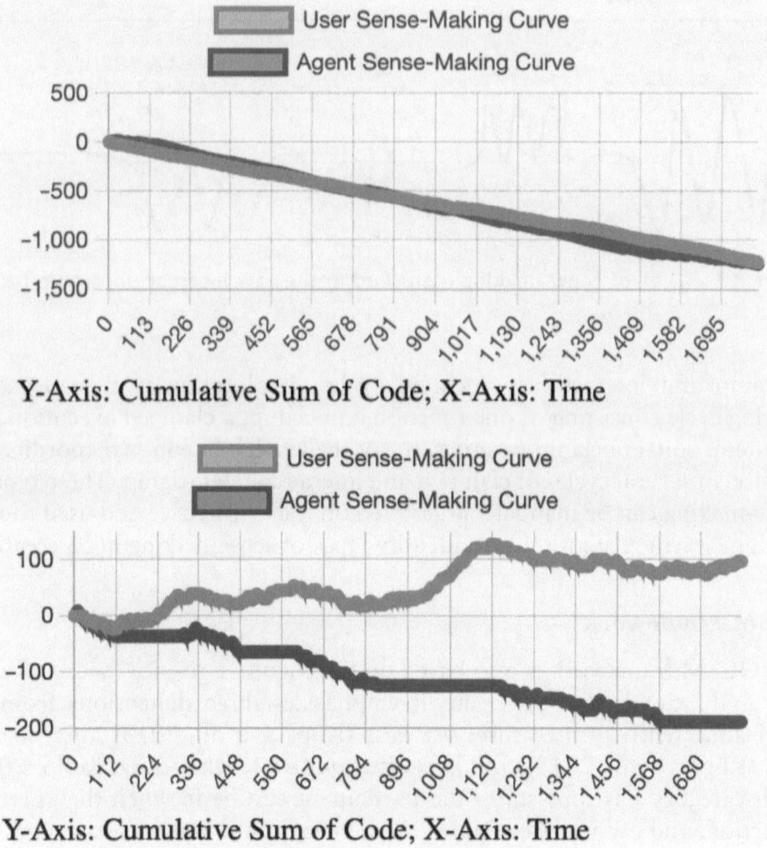

Figure 4.4 The CSM curve for a co-creative drawing session with an agent coded using the standard CSM coding procedure (left), and the CSM curve for a co-creative drawing session with an agent coded using the updated CSM coding procedure (right).

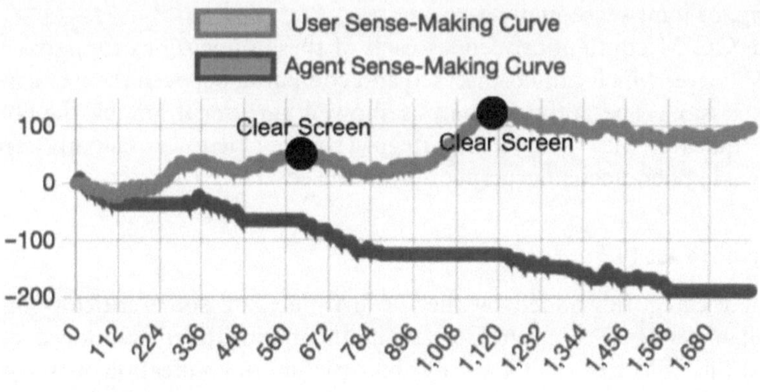

Figure 4.5 Mapping when the user cleared the screen to the CSM curve in a co-creative drawing session.

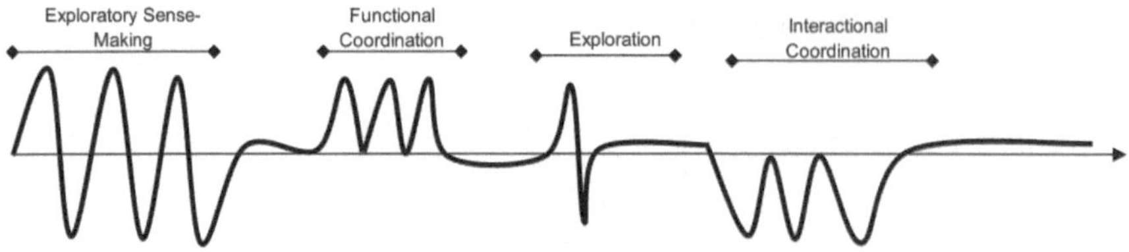

Figure 4.6 Types of sense-making identified in the raw interaction dynamic data.

unclamp and interactional unclamp cycles. Exploration is a single sense-making cycle consisting of one functional unclamp, a clamped execution, and an interactional unclamp (waiting) to see the results. Intentional coordination consists of several cycles of clamped and interactional unclamp. These types of sense-making can be manually annotated on the CSM curve and used to train machine learning algorithms to identify types of sense-making in co-creation.

OCSM Framework

The OCSM framework is also based on the cognitive science theory of enaction and extends the CSM, but it emphasizes three dimensions from the PSM and creativity literature: *newness* (Kupers et al., 2018), *appropriateness* (Kupers et al., 2018), and *participation* (De Jaegher & Di Paolo, 2007). Each category has four states the participant can be in when they perform an action, and each of these states has a behavioural marker associated with it. Newness relates to how novel the current action was, given the previous actions in the session. It can either be a repetition, minor modification, major modification, or new movement. Appropriateness relates to the value of the action and whether it was off-task, somewhat related, on-task, or explicit reference to task. The participation relates to the sense-making occurring, whether it was no participation, individual sense-making, reactive sense-making, or joint sense-making.

The OCSM continuously codes each of these dimensions to arrive at an OCSM curve, which can be analysed and compared between the participants of the session. The author's analysis showed agreement among the coders, and it indicated that the selected categories were able to consistently capture observable phenomena occurring during the co-creative session.

Extending CSM

The new categories added by the OCSM elucidate some critically important elements of co-creation. Combining them with the interaction dynamic data of the CSM can yield a more robust picture of co-creation, as shown in Figure 4.7.

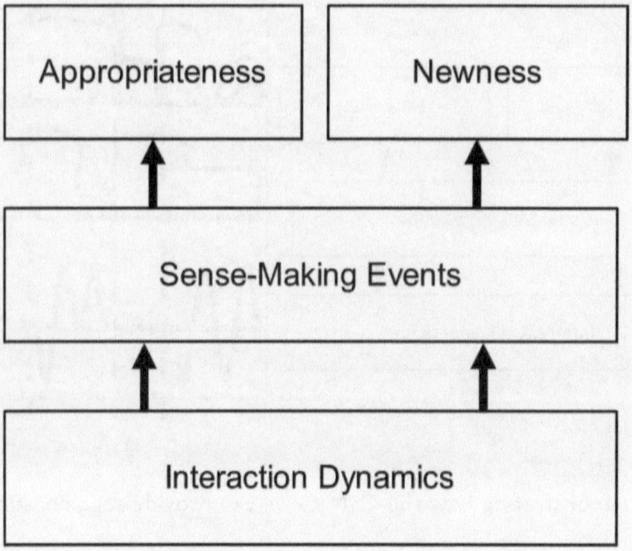

Figure 4.7 Extended CSM framework that integrates the new sense-making (participation), appropriateness, and newness categories.

The raw interaction dynamic data form sense-making events in Figure 4.8. For example, there may be exploratory sense-making that is individual, exploratory sense-making that is joint, or functional coordination that is either individual or joint. With the CSM data, it is possible to quantify how exploratory an event was, that is how off-task an action was. In this approach, the interaction dynamics inform the sense-making events, and the sense-making events each have a corresponding newness and appropriateness value.

When viewed in tandem, the CSM and OCSM generate a more full view of co-creation. The interaction dynamic data can provide points of interest to segment the data, as shown in Figure 4.8. Once the data are segmented, one can determine what OCSM codes were applied to that segment and investigate any correlational relationships that might arise. The CSM curve segments the data to classify events in the data, and the events can be used as a lens to view the data. There are several different events based on analysing the CSM curve. The segmentation is done based on time spent unclamped versus time executing fluid actions. The fluctuations are the sense-making process, and the clamped event is the action that was performed once the participant understood how the system in the world worked relative to their intentions. With this segmentation, it is possible to count the number of code applications of participation, newness, and appropriateness per segment to determine what was happening during each event.

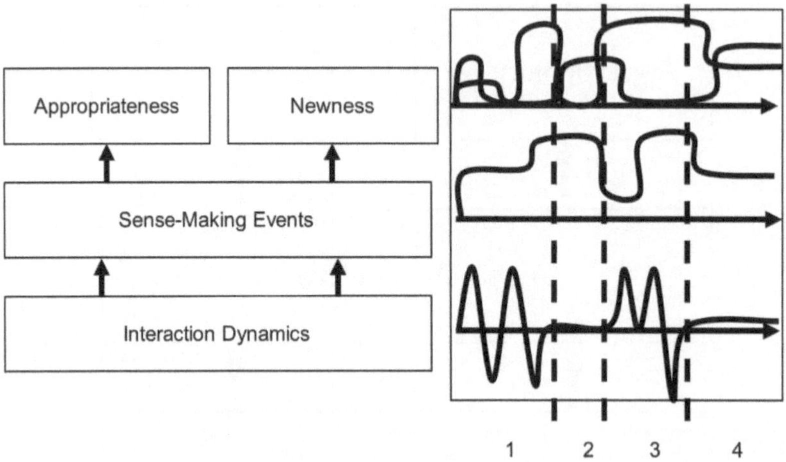

Figure 4.8 Demonstrating how the CSM curve can provide segmentation data.

Domain Applications

To help with the adoption of the CSM, the coding scheme has been developed for multiple co-creative contexts. To simplify the deployment of this coding procedure for each of the application domains, a paired-down version of the CSM coding method was used, which included four main categories of interaction: communicate, manipulate interface, execute, and wait, as shown in Table 4.1.

Discussion

CSM is a cognitive framework for modelling interaction dynamics in co-creative AI. This chapter simplified and updated the coding scheme of the CSM to include four modes of interaction: communicate with the AI agent, manipulate the interface of the software environment, fluidly execute creative actions, and wait. These interactions are coded continuously through time in the CSM framework to produce what is known as the CSM curve. The CSM curve illustrates the sense-making processes occurring in co-creation in a visually readable manner. The curve can be manually analysed and annotated, or it can be analysed with machine learning to identify trends and patterns. This chapter proposed several types of sense-making patterns based on different types of sense-making curves that could be generated by the CSM method. It also simplified and extended the CSM framework to account for new research in the field. CSM coding schemes were developed for several domains of co-creative AI to support research in co-creative AI. The CSM can be used to automatically code interaction dynamics in co-creative AI

Table 4.1 CSM coding scheme for different co-creative domains.

	Communicate	*Manipulate Interface*	*Execute*	*Wait*
Coded Value	**1**	**0.5**	**−1**	**0**
Drawing	Communicating to the agent about the drawing, e.g. instructing the agent what to do, positive/negative feedback	Using interface of tool, e.g. changing brush colour, selecting drawing modes (fill, erase, smudge)	Fluidly drawing	Pausing, hesitating
Pretend Play	Communicating to the agent about the play scenario, e.g. 'let's play cars'	Exploring play environment, e.g. searching for new toys, arranging the toy space	Executing an embodied play action	Pausing, hesitating
Dance	Communicating to the agent about the dance, e.g. demonstrating the dance task	Manipulating user interface of tool, e.g. selecting the dance style	Dance actions that are on-task	Pausing, hesitating
Music	Communicating with the agent about the musical improvisation e.g. Shimon's head bobbing recognition (Hoffman & Weinberg, 2010)	Interacting with the interface of the tool, e.g. setting the tempo, selecting a musical scale	Fluidly playing music	Pausing, hesitating
Design	Communicating to the agent about the design, e.g. describing a problem/solution, domain description	Interface actions, e.g. selecting utensils, arranging items on a design board	On-task sketching	Pausing, hesitating
Writing	Communicating to the agent about the written artefact, e.g. requesting assistance to generate or analyse text	Directly manipulating interface, e.g. scrolling the document, searching the document, changing font sizes	Fluidly typing/writing	Pausing, hesitating, looking away from the document
Programming	Communicating to the agent about the code, e.g. asking it to generate functions	Directly manipulating interface, e.g. scrolling through the program, using the search function	Fluidly typing out code	Pausing, hesitating

systems, enabling researchers to model interaction dynamics in new ways. Using the CSM to automatically code interactions occurring in co-creation can create a new type of co-creative system referred to as a *quantified co-creative AI system*, which automatically models the interaction dynamics of co-creation happening on the platform.

This chapter explained the value the CSM curve has in quantifying and modelling interaction dynamics. In particular, it described how the interaction dynamic data from the CSM are like the raw interaction data upon which the categories from the OCSM build. The raw interaction data can be used to classify different types of sense-making processes that occur during an improvisational co-creative session. The extended CSM was introduced that accounts for the new additions made by OCSM and shows how each component relates to each other. The extended CSM shows how interaction dynamics lead to sense-making events, and how each sense-making event has both an appropriateness value and a newness quality. The new framework offers a more holistic view of co-creation than either the CSM or OCSM could do alone.

By combining the CSM curve with the OCSM curves, it will be possible to effectively label the CSM curve for machine learning purposes. It is possible to find trends and patterns in the CSM curve and examine their relationship to the OCSM coding states. Then, it may be possible to automatically recognize some of the OCSM states from the raw sense-making data, such as the type of sense-making occurring and the appropriateness of the actions. Newness relates to the content of the action, rather than the dynamics of the action, so it cannot be automatically coded using the approach outlined here. It must be coded by either human observers or action recognition algorithms.

To integrate the CSM coding procedure into a co-creative system, there must be a timer that records the interaction state of both the user and the agent at specified time intervals (e.g. execute recordState every 0.5 seconds). The interaction state can be: communicate (coded as 1), manipulate interface (coded as 0.5), execute (coded as –1), and wait (coded as 0). While the user is communicating to the agent, a boolean variable (e.g. activeCommunication) should be set to true. While the user is clicking on elements of the interface, a boolean variable (e.g. activeInterface) should be set to true. While the user is executing actions on the shared creative product, there should be a boolean variable (e.g. activeExecute) set to true. If none of the previous conditions are true, then the user must be waiting. When the timer checks what state the user is in, these variables are used to determine what coding value to push to an array of user codes and agent codes.

Future Research Agenda

Quantifying the interaction dynamics of co-creation enables new types of analysis and evaluation. In particular, it elucidates differences in the co-creative process between individuals and among co-creative systems. The CSM enables a domain independent measure for investigating the interaction dynamics of co-creation, which can provide a common platform for researchers to compare the interactions of individuals as they engage in co-creation in different domains. This project has three research questions guiding future work.

RQ1: What types of sense-making processes (e.g. individual, reactive, or joint) and interaction dynamics (e.g. clamping/unclamping, turn-taking, interaction coupling, communication) emerge in co-creation between a user and co-creative AI agent?

This question explores how users make sense of the system through their interaction. For example, are there trends that characterize novice versus expert use of the co-creative system? This question can be utilized to evaluate the interaction dynamics of co-creative systems. Some systems may only facilitate individual or reactive sense-making, whereas others might achieve joint or PSM. Additionally, considering the interaction dynamics reveals insights about the temporal dimension of the co-creative process, such as turn-taking and interaction coupling.

RQ2: How might co-creative systems automatically model the interaction dynamics of co-creation happening in the system?

Future work can explore how to develop quantified co-creative AI systems that automatically detect sense-making and PSM patterns in co-creation. Identifying patterns in sense-making can be used for two purposes: (1) to guide the interactions of the co-creative agent, and (2) to aid in the analysis of co-creation. For example, the system could introduce novel ideas during functional coordination and semantically related ideas during interactional coordination. The sense-making patterns could enable two types of analysis in co-creative AI systems: (1) classifying the stages or phases of co-creation, and (2) classifying types of co-creative users.

Recent work by Davis (2023) developed the AI Drawing Partner, a co-creative drawing agent that serves as an experimental research platform to quantify and model the interaction dynamics of co-creation. The quantified co-creative AI system automatically captures and codes the user using the CSM framework based on how the user is interacting with the system. Capturing this information turns the co-creative AI system into a research platform to model interaction dynamics, as shown in Figure 4.9.

RQ3: How can co-creative agents leverage models of interaction dynamics to govern their creative behaviour?

Once a co-creative system is able to quantify and model interaction dynamics, it can utilize this model to influence its creative decisions. For example, the system could detect that the user is in an exploratory ideational mode of interaction and generate many new ideas to facilitate this exploration. Alternatively, the system could detect that the user is refining their creative product, and it could converge on one idea and deepen it. The idea is that the co-creative process is dynamic, with different types of interactions happening at different stages of the process. This question explores methods of recognizing high-level creative actions based on interaction dynamics and using this information to inform the creative actions of the AI.

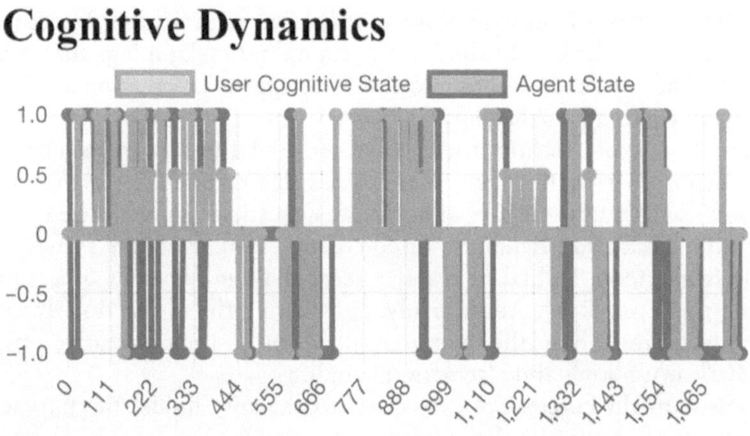

Execute = Drawing = -1
Interface Manipulation = 0.5
Communication = 1
Wait = No action = 0

Code	# of Applications
User Execute	272
User Communicate	244
User Interface Manipulation	97
User Wait	1154
User Average Value	0.01
Agent Execute	363
Agent Communicate	446
Agent Wait	958
Agent Average Value	0.05

Figure 4.9 Raw interaction data (left) and CSM curve (right) of a 15-minute drawing session captured and visualized by the AI Drawing Partner research platform.

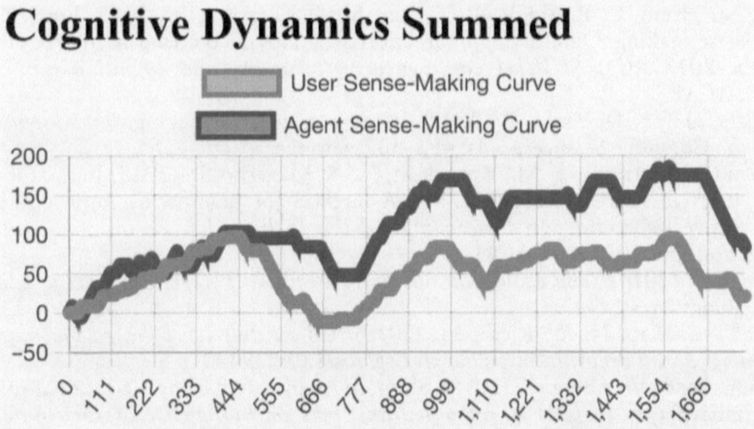

Y-Axis: Cumulative Sum of Code; X-Axis: Time

Analysis Type	Value
User Average Slope	0.02
User Y-Intercept	35.26
User R-Squared	0.0910
Agent Average Slope	0.08
Agent Y-Intercept	43.30
Agent R-Squared	0.6835
Pearson Coefficient	0.62

Figure 4.9 Continued

References

Biles, J. A. (2002). GenJam: Evolution of a jazz improviser. In *Creative evolutionary systems* (pp. 165–187). Morgan Kaufmann.

Csikszentmihalyi, M., Abuhamdeh, S., & Nakamura, J. (2014). Flow. *Flow and the Foundations of Positive Psychology: The Collected Works of Mihaly Csikszentmihalyi*, 227–238.

Davis, N. (2013). Human-computer co-creativity: Blending human and computational creativity. *Proceedings of the AAAI Conference on Artificial Intelligence and Interactive Digital Entertainment*, *9*(6), 9–12.

Davis, N. (2023). *AI drawing partner: Co-creative drawing agent and research platform to model interaction dynamics* [Manuscript in preparation].

Davis, N., Hsiao, C. P., Popova, Y., & Magerko, B. (2015). An enactive model of creativity for computational collaboration and co-creation. *Creativity in the Digital Age*, 109–133.

Davis, N., Hsiao, C. P., Singh, K. Y., Lin, B., & Magerko, B. (2017, June). Creative sense-making: Quantifying interaction dynamics in co-creation. In *Proceedings of the 2017 ACM SIGCHI conference on creativity and cognition* (pp. 356–366). ACM.

De Jaegher, H., & Di Paolo, E. (2007). Participatory sense-making. *Phenomenology and the Cognitive Sciences*, 6(4), 485–507. Springer.

Deshpande, M., Trajkova, M., Knowlton, A., & Magerko, B. (2023, June). Observable creative sense-making (OCSM): A method for quantifying improvisational co-creative interaction. In *Proceedings of the 15th conference on creativity and cognition* (pp. 103–115). ACM.

Gibson, J. J. (2014). *The ecological approach to visual perception: Classic edition.* Psychology Press.

Guzdial, M., Liao, N., & Riedl, M. (2018). *Co-creative level design via machine learning. arXiv Preprint.* https://arxiv.org/abs/1809.09420

Hoffman, G., & Weinberg, G. (2010). Shimon: An interactive improvisational robotic marimba player. In *CHI'10 extended abstracts on human factors in computing systems* (pp. 3097–3102). CHI.

Ibarrola, F., Lawton, T., & Grace, K. (2022). *A collaborative, interactive and context-aware drawing agent for co-creative design. arXiv Preprint.* https://arxiv.org/abs/2209.12588

Kantosalo, A., & Riihiaho, S. (2019). Usability testing and feedback collection in a school context: Case poetry machine. *Ergonomics in Design*, 27(3), 17–23.

Kantosalo, A., & Takala, T. (2020, September). *Five C's for human-computer co-creativity-an update on classical creativity perspectives* (pp. 17–24). ICCC.

Kantosalo, A., Toivanen, J. M., Xiao, P., & Toivonen, H. (2014). *From isolation to involvement: Adapting machine creativity software to support human-computer co-creation* (pp. 1–7). Proceedings of the Fifth International Conference on Computational Creativity, Ljubljana, Slovenia.

Karimi, P., Maher, M. L., Davis, N., & Grace, K. (2019). *Deep learning in a computational model for conceptual shifts in a co-creative design system. arXiv Preprint.* https://arxiv.org/abs/1906.10188

Kellas, J. K., & Trees, A. R. (2014). Rating interactional sense-making in the process of joint storytelling. In *The sourcebook of nonverbal measures* (pp. 281–294). Psychology Press.

Kupers, E., Van Dijk, M., & Lehmann-Wermser, A. (2018). Creativity in the here and now: A generic, micro-developmental measure of creativity. *Frontiers in Psychology*, *9*, 2095. Frontiers Media SA.

Magerko, B., DeLeon, C., & Dohogne, P. (2011, September 15–17). *Digital improvisational theatre: Party quirks* (pp. 42–47). Intelligent Virtual Agents: 10th International Conference, IVA 2011, Reykjavik, Iceland, Proceedings 11, Springer, Berlin, Heidelberg.

Norman, D. A. (1999). Affordance, conventions, and design. *Interactions*, *6*(3), 38–43.

Norman, D. A. (2013). *The design of everyday things* (revised and expanded ed.). Basic Books.

Rezwana, J., & Maher, M. L. (2022). *Designing creative AI partners with COFI: A framework for modeling interaction in human-AI co-creative systems.* ACM Transactions on Computer-Human Interaction.

Varela, F. J., Thompson, E., & Rosch, E. (2017). *The embodied mind, revised edition: Cognitive science and human experience.* MIT Press.

Yannakakis, G. N., Liapis, A., & Alexopoulos, C. (2014). *Mixed-initiative co-creativity.* Proceedings of the 9th International Conference on the Foundations of Digital Games, FDG 2014. https://antoniosliapis.com/papers/mixed_initiative_co-creativity.pdf

Chapter 5

Demystifying AI Techniques for Creative Domains

A Beginner's Guide to Empowering Human Creativity

Thorsten Fröhlich

Introduction

Defining *creativity* has been a research focus for many years, resulting in various variants. Walia (2019) summarised that focusing solely on the result (creation) can disservice the entire creative process because it neglects the creative process of equal importance. Instead, *creativity* should be studied flexibly, understanding the dynamics and interplay of various forces, including emotional or mental constructs. Creation results from many variables. So, *creativity* cannot be understood merely by achievements and successes. *Creativity* is expressed through integrating elements like person, process, products, persuasion, interactions, and the idea that creative potential may or may not result in tangible creation.

That is where artificial intelligence (AI) fits in. AI techniques are applied across creative domains like the arts, sciences, business, and education to enhance or mimic human creativity.

- Creativity is often synonymous with originality and aesthetic innovation in the arts, giving rise to new forms, styles, or interpretations in visual arts, music, literature, and performance.
- In the sciences, creativity fuels groundbreaking research and discoveries, as scientists think beyond conventional wisdom to develop new theories, methods, or technologies.
- Business creativity involves strategically generating new ideas for products, services, or processes, enhancing competitive advantage and driving growth.
- Finally, creativity fosters a dynamic learning environment in education where educators and students engage in critical thinking, problem-solving, and cultivating a lifelong love for Learning.

Together, these fields showcase creativity's diverse and vital role in shaping human thought and endeavour across disciplines.

DOI: 10.4324/9781003453901-7

Understanding the intricacies of AI technologies can be a challenge for newcomers in AI-augmented creativity. Providing a glossary elucidating AI techniques using metaphors and real-world applications offers an accessible entry point for these individuals. Such a glossary serves as an educational scaffold, allowing newcomers to grasp complex algorithms and their utility in creative tasks without requiring an in-depth technical background. Additionally, a glossary can facilitate interdisciplinary dialogue by establishing a common vocabulary between technologists and creatives. This foundational understanding may empower individuals to harness AI tools in their creative work better, fostering innovation and possibly leading to groundbreaking advancements in creative fields. Therefore, a glossary demystifying AI techniques is not merely a didactic instrument but also an enabler of cross-disciplinary collaboration and creative exploration.

The glossary presented focuses solely on the most salient terms and techniques concerning creativity in AI. It is not an exhaustive compilation and intentionally omits less central concepts to facilitate a streamlined introduction to the subject matter. The selection of terms has been curated to offer a foundational understanding beneficial to newcomers and interdisciplinary practitioners in AI-augmented creativity.

The glossary is organised as follows:

Table 5.1 Glossary of terms.

Field	*Subtopic*	*Specialisation*
Machine learning		
	Supervised learning	
		• Supervised learning with linear regression
		• Supervised learning with support vector machines (SVMs)
		• Supervised learning with decision trees
		• Supervised learning with random forests
	Unsupervised learning	
	Reinforcement learning	
Deep learning		
	Artificial neural networks (ANNs)	
	Convolutional neural networks (CNNs)	
	Recurrent neural networks (RNNs)	
	Supervised learning with neural networks	
	Generative models	
		• Generative adversarial networks (GANs)
	Transformers	
		• Generative pre-trained transformer (GPT)
		• Bidirectional encoder representations from transformers (BERTs)
Natural language processing (NLP)		
	Text understanding	
	Sentiment analysis	

(*Continued*)

Table 5.1 (Continued)

Field	*Subtopic*	*Specialisation*
	Machine translation	
	Text generation	
Computer vision		
	Image classification	
	Object detection	
	Style transfer	
Computer hearing		
	Voice and speech recognition	
Evolutionary algorithms		
	Genetic algorithms	
	Differential evolution	
	Particle swarm optimisation	

Glossary of AI Techniques for Creative Tasks

AI offers an array of techniques that can be applied to a myriad of creative tasks. Next, the hierarchy of such methods is categorised under broader terms to enable better understanding.

Machine Learning

Machine learning is akin to an artist's palette brimming with a multitude of colours, techniques, and possibilities. Machine learning algorithms adapt and evolve just as an artist mixes colours to create new hues or uses brushstrokes to express different emotions. They draw from vast data pools, learning to identify patterns and make decisions that improve over time (Zhou, 2021).

Supervised Learning

Supervised learning is comparable to an apprentice sculptor carefully chiselling away at a block of marble under the watchful eye of a master. The master references a model sculpture, allowing the apprentice to understand how the final piece should look.

In this paradigm, algorithms learn from labelled training data, making predictions based on known input–output pairs. The labels serve as the master sculptor's model, guiding the algorithm towards the intended result. The algorithm is 'supervised' as it learns, receiving feedback and making adjustments until its predictions closely align with the desired outcomes.

In music, supervised learning can be used for genre classification. By training the algorithm on labelled samples of different genres, it learns to classify new, unlabelled music into one of the known genres.

Supervised Learning with Linear Regression

Supervised learning with linear regression can be likened to fitting a straight line through a series of points on a graph, similar to connecting the dots in a child's activity book. Imagine each point on the graph represents the relationship between studying hours and exam scores for different students. The aim is to draw a line that best describes this relationship, much like a child aims to connect dots to reveal a picture accurately.

In the supervised learning context, the 'dots' are the labelled data points, and the 'line' is the mathematical model learned by linear regression. Just as the line seeks to pass as closely as possible through all the dots, linear regression aims to find the equation of the line that minimises the distance between itself and each data point. Once this line is established, it can be used to predict a student's exam score based on the number of hours they've studied – akin to how a child can anticipate the picture's shape even before all dots are connected.

For instance, in content recommendation systems, linear regression could predict the appeal of a new piece of art based on features like the colour scheme, size, and medium.

Supervised Learning with Support Vector Machines (SVM)

Supervised learning using SVM can be compared to a game of dodgeball. Imagine two teams on opposite sides of a court, each representing a different category in a classification problem. SVM aims to find the widest 'safety zone' or margin where no player from either team can cross; this is the optimal dividing line between the two classes.

In this metaphor, the 'support vectors' are akin to the key players standing closest to this dividing line, as they are the ones who essentially define its position. By identifying the widest possible safety zone – maximising the margin – the SVM algorithm ensures that new players (or data points) entering the court can be accurately categorised into one of the two teams, just as you would instantly know which side of the court a new player belongs to based on where they stand relative to the dividing line.

Thus, in a similar way that the dividing line separates the dodgeball teams and predicts where new players belong, SVM seeks to find the hyperplane that best divides the data points of different classes in a multidimensional space and accurately classifies new, unseen data.

SVM could be used to auto-tag photos with artistic styles like 'Impressionist', 'Cubist', or 'Surrealist'.

Supervised Learning with Decision Trees

Supervised learning using decision trees can be likened to navigating a maze. Each decision point in the labyrinth corresponds to a node in the decision tree. You must choose which path to take at each decision point based on specific criteria – much like how a decision tree uses data features to decide the next node to traverse. These criteria could be as simple as 'turn left if you see a red flag, and right if you see a blue flag'.

As you navigate the maze using these rules (analogous to the 'if-then' conditions in a decision tree), you arrive at an exit point corresponding to a leaf node that gives you the final classification or prediction. If you've navigated successfully, you've found the most efficient path through the maze – similar to how a well-trained decision tree would accurately classify or predict the target variable.

Decision trees use a hierarchical, top-down approach to make decisions, guiding data down the right 'path' based on feature values until a prediction or classification is reached.

In scriptwriting, a decision tree could recommend plot directions based on previous successful story arcs.

Supervised Learning with Random Forests

Supervised learning with random forests can be compared to a democratic voting system in a small town deciding on a new community project. Imagine that instead of a single mayor making the decision (akin to a single decision tree), the city relies on a council of diverse experts (ensemble of decision trees). Each expert specialises in a particular area – healthcare, recreation, infrastructure, and so on.

When a new project proposal comes up, such as building a new park, each expert evaluates the idea based on their speciality and criteria. Some might assess soil quality, others the potential for community engagement, and so forth. Individually, their judgements may be prone to biases or limitations of their specific expertise, much like individual decision trees can overfit their training data.

After all experts have conducted their evaluations, they cast their votes. The majority vote is considered the final decision for the project, just as the random forest algorithm takes the mode (for classification) or mean (for regression) of all the decisions made by individual trees in the ensemble.

The strength of random forests is in aggregating diverse perspectives (decision trees trained on different subsets of features and data) to arrive at a more balanced and reliable prediction or classification.

In music composition, a random forest could generate a piece that incorporates multiple styles or genres, predicting the likelihood of audience acceptance.

Supervised Learning with Neural Networks

Supervised learning using neural networks can be likened to an intricate assembly line in a factory designed to manufacture complex products like smartphones. Each station on the assembly line corresponds to a node in a neural network layer, and the whole assembly line itself can be considered the network architecture.

As raw materials (input data) move through the assembly line, each station performs specific tasks, such as installing a chip or assembling the screen. This mirrors the weighted sum and activation function calculations at each node. As the product moves from one station to the next, it gets more refined and closer to the final product – akin to how data are transformed as it moves through layers in a neural network.

Ultimately, the fully assembled smartphone (the output) must meet quality control standards (target variable). If it does not, feedback is provided to optimise the assembly line's performance, like how backpropagation adjusts the weights in a neural network during training.

A neural network can capture intricate relationships in data through its layers and nodes, transforming raw input into a meaningful output, all the while continuously refining its performance through feedback mechanisms.

In the realm of digital art, neural networks can generate highly detailed and complex pieces, or in NLP, they can help in writing poems or stories.

Unsupervised Learning

Picture a jazz musician freely improvising on stage. There is no set list or predefined structure; the musician explores the musical space, creating melody and harmony on the fly.

Unsupervised learning algorithms operate on unlabelled data, finding structure and patterns on their own. Like a jazz musician navigating scales and chords without a set list, the algorithm clusters or segregates data based on inherent characteristics without having a predetermined 'correct' answer to guide it.

In visual arts, unsupervised learning techniques can be used for image segmentation, helping artists and designers automatically separate elements within a complex scene for further editing or analysis.

Reinforcement Learning

Think of a playwright who constantly revises a script based on audience reactions. Applause, laughter, or a standing ovation reinforces certain lines or scenes, prompting the playwright to keep them in future iterations.

Reinforcement learning centres on an agent that performs actions within an environment to maximise a reward function. Like the playwright refining a script based on audience feedback, the algorithm receives rewards or penalties for its actions, thus learning how to achieve the best possible outcome over time (Sutton & Barto, 2020).

In video game design, reinforcement learning can train non-player characters (NPCs) to interact with human players in more complex, engaging ways. The NPCs adapt based on the actions and strategies of human players, leading to a more dynamic gaming experience.

Deep Learning

Deep learning is like the work of a master storyteller, weaving complex narratives with intricate layers and subtexts. Just as stories can have plots, sub-plots, and character arcs that interact to create a captivating tale, deep learning algorithms utilise multiple layers of artificial neurons to extract increasingly abstract features from raw data. These layers are organised hierarchically within neural networks, which serve as the scaffolding for deep learning models.

In neural networks, the basic building blocks are artificial neurons inspired by biological neurons in the human brain. These artificial neurons receive input, process it, and pass the output to the next layer. Each neuron's output is a simple function of its input, often passed through a non-linear activation function to introduce complexity and enable learning of non-linear patterns (Skansi, 2018).

When organised into layers, these neurons can handle increasingly complicated tasks. The first layer, typically known as the input layer, receives the raw data. This is analogous to the exposition in a story, where all the characters and settings are introduced. Subsequent layers, often termed hidden layers, transform the raw data into higher-level features. These are like the developing events in a story, contributing more complexity and details. The final layer, or the output layer, makes the ultimate prediction or classification, similar to the resolution of a story where all the plot lines converge and climax.

The strength of the connections between neurons, known as weights, is adjusted during the learning process. This is done using optimisation algorithms like gradient descent, which iteratively tweak these weights to minimise a loss function. The loss function acts like an editor, providing feedback to the neural network on how well it's doing, enabling it to refine its 'narrative' for better accuracy.

In practical applications, deep learning has excelled in fields such as image recognition, NLP, and game playing. For example, CNNs have been particularly effective in image classification tasks, where they can accurately identify objects within an image. Similarly, transformer-based models like GPT and BERT have set new benchmarks in natural language understanding and generation, capable of tasks ranging from machine translation to summarisation. Like a master storyteller, deep learning takes the rudimentary elements of data and crafts them into meaningful and often insightful outcomes.

Artificial Neural Networks (ANN)

An ANN is like an orchestra, where each musician plays a specific instrument. They listen to each other and the conductor, harmonising to produce a melodious output.

ANNs consist of interconnected nodes, or 'neurons', arranged in layers. Data flow through these layers, being transformed at each node, much like how each musician adds complexity to the overall melody. This allows ANNs to perform various tasks, from simple linear classifications to complex functions.

In literature, ANNs have been used to generate poetry or even short stories, mimicking human-like creativity in crafting narratives.

Convolutional Neural Networks

A CNN is akin to a film editor who meticulously focuses on each frame, recognising patterns like movement and light to create a seamless visual story.

CNNs are specialised for processing grid-like data structures, most commonly images. They employ a mathematical operation called convolution to focus on local regions of the input data, effectively enabling the network to recognise patterns hierarchically.

In graphic design, CNNs are often used for image enhancement, automatically improving visual quality and aesthetic appeal.

Recurrent Neural Networks

An RNN resembles a playwright who recalls previous dialogues and events in a drama to shape the next act.

RNNs have loops to allow information persistence, meaning they remember past information to inform the processing of current inputs. This memory feature makes them ideal for sequence-related tasks.

In music composition, RNNs can be used to generate melodies that have a sense of temporal coherence, creating a musical piece that flows logically from start to finish.

Generative Models

Imagine a skilled mimic who can imitate famous personalities and come up with new sentences in their distinctive styles.

Generative models aim to learn the underlying distribution of the data to generate new, similar data points. They capture the essence of the input data, much like a mimic captures the essence of the personality they are imitating.

In visual arts, generative models are used to create new artwork that mimics the style of existing artists, expanding the boundaries of creative expression.

GENERATIVE ADVERSARIAL NETWORKS

GANs have brought a fresh perspective to the generative models in deep learning. They are like the classic duo of a painter and a critic, where the painter tries to create a masterpiece, and the critic evaluates it, pushing the painter to improve continuously.

GANs are similar to a dance-off between two talented artists. One artist (the Generator) tries to perform a routine so convincing that the other artist (the Discriminator) can't tell if it's improvised or choreographed.

GANs consist of two neural networks: the Generator and the Discriminator. The Generator produces data that it tries to make indistinguishable from real data. The Discriminator evaluates the data it receives and distinguishes between real and fake. This adversarial process leads to the Generator improving its ability to create realistic data, akin to how dancers in a dance-off push each other to excel (Kaddoura, 2023).

Artists are using GANs to create artworks auctioned as unique pieces, pushing the boundaries of what is considered art. Filmmakers have employed GANs to upscale vintage films, improving their quality while retaining their original essence.

Transformers

A transformer model is like a detective who can simultaneously listen to multiple witnesses, cross-reference their statements, and draw insightful conclusions about a complex case.

Transformers are designed to handle sequences, but unlike RNNs, they can consider all sequence positions simultaneously, allowing for more parallelisation and efficient handling of long sequences.

In journalism, transformer models can assist in summarising long articles or even generating drafts, allowing writers to focus more on nuanced storytelling and less on data gathering and basic content creation.

GENERATIVE PRE-TRAINED TRANSFORMER

The GPT is an exceptional addition to the deep learning domain. It is like a Swiss Army knife for language tasks. GPT models excel at generating human-like text and are pre-trained on a massive corpus, making them highly flexible and capable in a wide array of language tasks.

GPT is akin to a versatile playwright who is proficient in writing scripts, dialogues, monologues, and even poetry. The model has a broad understanding of different styles and tones and uses this knowledge to generate text that seamlessly fits into various contexts.

GPT employs the transformer architecture, specifically focusing on the decoder part of it. When generating the next word, it can use attention mechanisms to consider the context of all the previous words in a sequence. Like a playwright who understands the flow of dialogue and the mood of a scene, GPT synthesises these elements to create coherent, contextually relevant, and sometimes even captivating text.

GPT can generate dialogues for a play or plot outlines, serving as a creative aid to scriptwriters. Artists have used GPT models to co-write poems, blending machine-generated creativity with human artistry.

BIDIRECTIONAL ENCODER REPRESENTATIONS FROM TRANSFORMERS

BERT stands as a revolutionary force in the world of natural language understanding. It is like an investigative journalist who digs deep into the text, understanding the nuances from multiple directions to get the complete picture.

Imagine BERT as an archaeologist, delicately sifting through text layers to understand the relationship between words and their contextual meanings.

Unlike traditional language models that read text from left to right or right to left, BERT is designed to read text bi-directionally. This allows it to capture the context from both sides of a word, similar to how an archaeologist would study the layer above and below an artefact to understand its historical context.

BERT algorithms can filter through many articles or art pieces, identifying the most contextually relevant works for a thematic exhibition.

BERT can be used to develop more nuanced and context-aware dialogue systems for interactive storytelling platforms, elevating the user experience.

Natural Language Processing

NLP is like a skilled linguist proficient in multiple languages, dialects, and cultural nuances. As linguists decipher meaning, context, and sentiment in human interactions, NLP algorithms analyse, understand, and generate human language (Chowdhary, 2020).

Text Understanding

Consider text understanding as a literature professor dissecting a novel, identifying themes, characters, and narrative techniques to extract the core message.

Text understanding in NLP involves various sub-tasks like tokenisation, part-of-speech tagging, and semantic analysis. These steps allow the algorithm to break down text data and interpret its meaning, akin to how a literature professor analyses a novel.

In marketing, text understanding extracts insights from customer reviews, helping brands understand consumer sentiment and preferences.

Sentiment Analysis

Imagine sentiment analysis as a film critic who not only reviews a movie's technical aspects but also gauges the emotional impact on the audience.

Sentiment analysis categorises the emotional tone behind words, utilising text analytics, language modelling, and computational linguistics. It's like a critic who can dissect various aspects of a movie to gauge its overall emotional impact.

In social sciences, sentiment analysis can evaluate public opinions on sensitive issues, offering a nuanced understanding that quantitative data alone cannot provide.

Machine Translation

Picture a seasoned translator who can effortlessly convert a Russian poem into English, maintaining its literal meaning and lyrical beauty.

Machine translation involves converting text from one language to another. It's not just a word-for-word substitution but also consists of understanding context, idioms, and linguistic nuances to produce a translation that captures the original text's essence.

In publishing, machine translation is increasingly used to translate books into multiple languages, making them accessible to a global audience.

Text Generation

Text generation is like a jazz musician taking a standard tune and improvising to create something new yet familiar.

Text generation algorithms, often based on models like GPT or transformers, create new text based on the data they have been trained on. They can produce coherent and contextually relevant text over sequences akin to a jazz musician improvising on a standard tune.

In screenwriting, text generation algorithms can help generate dialogues or even entire scenes, serving as a creative aid for writers.

Computer Vision

Computer vision is like a world-class photographer with an analytical mind, capturing and interpreting scenes for their aesthetic value and compositional elements, such as objects, characters, and actions (Szeliski, 2022).

Image Classification

Picture a museum curator who sorts paintings into genres like Renaissance, Impressionism, and Abstract based on their features.

Image classification algorithms categorise images into predefined classes based on their features. They analyse pixels and patterns like how a curator would examine brush strokes and colour schemes.

In fashion design, image classification can be used to sort through different styles, patterns, or seasonal collections, automating what would be a labour-intensive process.

Object Detection

Imagine a wildlife photographer with an uncanny ability to spot hidden animals in a dense forest, capturing them in the frame without disturbing the environment.

Object detection algorithms identify and locate objects within images. They do this by analysing different portions of the image to identify shapes, features, and objects, much like a keen-eyed photographer would spot animals in the wild.

In automotive design, object detection is critical for developing autonomous vehicles, where the technology needs to recognise obstacles, traffic signals, and pedestrians.

Style Transfer

Think of style transfer as a talented mimic actor who can perform Shakespeare in the style of a modern-day sitcom.

Style transfer algorithms take the content from one image and the style from another to generate a new image. They separate and recombine features and styles, much like how an actor might adopt one character's body language and another's dialogue style.

In interior design, style transfer can help clients visualise how different styles, like Art Deco or Minimalist, would look in their living spaces.

Computer Hearing

Computer hearing is akin to a sound engineer with a highly trained ear, capable of isolating specific instruments in an orchestra, identifying audio anomalies and enhancing sound quality for the optimal listening experience.

Voice and Speech Recognition

Imagine a seasoned radio host who can instantly recognise callers based on their voice and manner of speaking.

Voice and speech recognition algorithms analyse audio signals to convert spoken words into text and identify individual voices. They dissect sound frequencies and patterns, similar to how radio hosts distinguish callers.

In podcasting, voice and speech recognition can be used for real-time transcription services, making the content more accessible to a broader audience.

Evolutionary Algorithms

Evolutionary algorithms are like Mother Nature's trial-and-error evolution method, albeit accelerated. As species adapt to their environments over generations, these algorithms modify and refine solutions over numerous iterations to solve complex problems (Bartz-Beielstein et al., 2014).

Genetic Algorithms

Think of genetic algorithms as dog breeders trying to create a new breed that combines the best traits of different existing species.

Genetic algorithms optimise solutions by mimicking the natural selection process. They start with a population of solutions and combine, mutate, and select them over generations to find the most optimal answer, much like how breeders would selectively mate dogs to achieve specific traits.

Genetic algorithms can optimise layouts and structures in architectural design, considering multiple variables like light, space, and materials.

Differential Evolution

Imagine a group of chefs tweaking a recipe. Each one changes an ingredient or a step, and they compare the outcomes to settle on the best version of the dish.

Differential evolution is a specific evolutionary algorithm used for optimisation. It iteratively refines candidate solutions by combining elements of different solutions and selecting the ones that improve the objective function, similar to chefs fine-tuning a recipe.

In the film industry, differential evolution can be used for optimising rendering algorithms, reducing the computational cost while maintaining high visual quality.

Particle Swarm Optimisation

Consider a flock of birds searching for food. They communicate and adjust their paths based on where each bird finds the most food, collectively arriving at the best foraging spot.

The social behaviour of birds and fish inspires particle swarm optimisation. Each 'particle' represents a potential solution and adjusts its position in the solution space based on its own experience and that of its neighbours, similar to birds converging on an optimal foraging spot.

In music production, particle swarm optimisation can be used to automatically tune the parameters of audio effects, saving time, and potentially discovering novel settings that a human might not consider.

Future Research Agenda

As these technologies continue to advance, some key questions emerge that warrant further research:

RQ1: How will fully autonomous AI creativity impact perceptions of art, science, and other creative works?

As AI can independently generate creative outputs without human involvement, studying how this affects our valuation and appreciation of creativity will be necessary.

RQ2: What are the ethical implications of AI augmenting or replacing human creativity?

As AI takes on more creative responsibilities, ethical concerns on copyright, attribution, and the potential displacement of human creators merit philosophical and legal analysis.

RQ3: How can we design AI creativity tools that optimise collaborative human–AI partnerships?

Rather than having AI work in isolation, there is great potential in developing AI co-creators that synergise with humans throughout the creative process. Research is needed to design these collaborative systems.

RQ4: What psychological effects will emerging creative AIs have on human identity and sense of creativity?

AI produces creative works indistinguishable from humans, which may impact human self-perception and motivation. Psychologists must investigate this evolving relationship between humans and creative AI.

RQ5: How can we build AI creativity benchmarks and evaluation frameworks that align with human measures?

Current AI benchmarks focus on technical prowess, but human creativity is multifaceted. Developing holistic benchmarks is critical for nurturing AI creativity that complements human values.

References

Bartz-Beielstein, T., Branke, J., Mehnen, J., & Mersmann, O. (2014). Evolutionary algorithms. *Wiley Interdisciplinary Reviews: Data Mining and Knowledge Discovery*, *4*(3), 178–195.

Chowdhary, K. (2020). Natural language processing. In K. Chowdhary (Ed.), *Fundamentals of artificial intelligence* (pp. 603–649). Springer.

Kaddoura, S. (2023). *A primer on generative adversarial networks*. Springer Nature.

Skansi, S. (2018). *Introduction to deep learning: From logical calculus to artificial intelligence*. Springer International Publishing.

Sutton, R., & Barto, A. G. (2020). *Reinforcement learning: An introduction*. The MIT Press.

Szeliski, R. (2022). *Computer vision* (2nd ed.). Springer Nature.

Walia, C. (2019). A dynamic definition of creativity. *Creativity Research Journal*, *31*(3), 237–247.

Zhou, Z.-H. (2021). *Machine learning*. Springer Nature.

Part 2

Critical Issues on Artificial Co-Creation

Chapter 6

Creating Images with Generative AI

An Imaginative Aid?

Claire Anscomb

Introduction

It is standardly held that agential creativity concerns the production of artefacts or states, such as artworks, that are both novel and valuable (Boden, 2004). Given that imagination is unlike other intentional states in that it is non-truth-bound, Stokes (2014) has proposed that the cognitive play imagination can facilitate may be '*essential to creative art-making for the simple reason that creative things are, in part, new things*' (pp. 158–159). Importantly, while traditional views have pictured artists as divinely inspired, attention to the art-making process reveals that even 'geniuses' like Picasso partake in a creative process that, in large part, consists of careful and deliberate thought (Stokes, 2014). As Gaut (2003) has proposed, active creativity seems more common and important in the arts – while a painter may suddenly 'see' how their painting will look:

> *[M]uch of the subsequent work will involve scrutinising the painting as it is being made, imagining how it could be improved by altering it in various ways, trying out these changes, observing the results, making more alterations, and so forth. And this process may take the painting far away from its original imagined look.*
>
> (p. 157)

Historically, these intertwined stages of imagining and making have all been carried out by human agents; however, technological developments have changed this.

With the invention of photography, for example, came a way of producing images, using a mind-independent process. However, the objects of photographically produced images are confined to those of the real world, given the core process of registering patterns of light, reflected from objects, on photosensitive surfaces (Anscomb, 2021). This is not true of more recent technologies involving machine learning (ML), such as generative adversarial networks (GANs), which can be used to produce images. Some artists (as

DOI: 10.4324/9781003453901-9

section "Creativity and Constraint" will examine) have used such technologies to facilitate the production of images and assist their imaginative powers. As artist Sarah Meyohas enthuses, algorithmic models can generate – '*images one couldn't have imagined*' (quoted in Ploin et al., 2022, p. 42).

Generating Images with Artificial Intelligence

Some artists like Sarah Meyohas, including Mario Klingermann, Robbie Barrat, Jake Elwes, Helena Sarin, David Young, and Anna Ridler, work very directly with GANs – to the extent that the boundaries between software engineering and art practice become porous. Reporting in 2022, Ploin et al. found, based on the practices of such artists, 'five new activities associated with the use of ML models in artistic practice: (1) technical research, (2) selecting or building models, (3) building datasets, (4) training models, and (5) curating outputs' (p. 21).

With regard to (2), as Helena Sarin has explained,

> *everybody who works seriously with GANs writes their own code, to different degrees. [. . .] I use CycleGAN but I still write a lot of code for pre-processing, for post-processing, for changing high-level parameters to automate some parts of the pipeline. That's what adds uniqueness to the art.*
>
> (quoted in Ploin et al., 2022, p. 22)

To gain even more control, Helena Sarin also, in relation to (4), combines models using techniques such as 'GAN-chaining', where outputs may be upscaled and passed through different GANs. Furthermore, as per (3), many of the aforementioned artists choose to experiment with existing, curated, and custom-made visual datasets (Ploin et al., 2022). Anna Ridler, for example, used 200 handmade ink drawings as her dataset to generate images for the 2017 work *The Fall of the House of Usher*, while Jake Elwes's production process for the 2019 work *Zizi* involved '*selecting a network pre-trained on celebrity faces and fine-tuning it with a hand-curated dataset*' consisting of a few thousand high-resolution images of drag performers to '*sort of corrupt, or dirty, their dataset*' (quoted in Ploin et al., 2022, p. 23). In doing so, both these artists created work that commented on the technology, from its uncanny qualities to queer erasure and normative visual grammars.

Notably, ML models function best with large amounts of data, so the choice to use relatively small training sets tends to result in '*machine failure aesthetics*' (Elgammal, 2023), or mislearning that many artists have found appealing (Ploin et al., 2022). Significantly, once artists have a fine-grain understanding of models, they can make artistically valuable creative choices, as Anna Ridler has explained: '*Each of these different models will have different associations and contexts that you can then start to play with*' (quoted in Ploin et al., 2022, p. 55). By making decisions based on such understanding, artists can enter into productive feedback loops with GANs and build on the surprising variations of existing images output by the models (Ploin et al., 2022). The nature and scope of creative choices massively changes however, with text-to-image generators, which have exploded in sophistication and popularity since Ploin et al.'s (2022) report was written.

By entering a text prompt, systems like DALL·E 2, Midjourney, and Stable Diffusion enable users to produce a range of images from the impressively photorealistic to highly stylized. As with other generative models, the case has been made that these artificial intelligence (AI) image generators enable users to do more than they imagined (Coeckelbergh, 2023) and play an important role when a user is unable to imagine how an image corresponding to their prompt could look (Khosrowi et al., 2023). Although users are largely restricted to entering text prompts, as Khosrowi et al. (2023) highlight, there are users who potentially can, and do, enter into back-and-forth interactions with systems like Stable Diffusion, and in doing so, exhibit far greater control, which may be further increased with the use of technologies like inpainting and ControlNet. However, this is still largely reliant on the ability to craft good prompts and, as Heaven (2022) has explained, prompts for the most detailed, stylized images can be hundreds of words long and so skilful is their production that online marketplaces have sprung up for such valuable assets.

Artist Boris Eldagsen, who controversially won a Sony world photography award with an AI-generated image has referred to what he does as '*promptography*'. He has outlined 11 parts of the prompt to craft, each of which gives scope for a range of choices that needn't be restricted by material boundaries and budgets, but he cautions: '*[T]wo thirds of the prompts are only good if you have knowledge and skills, when you know how photography works, when you know art history*' (quoted in Williams, 2023). Indeed, as Tosca (2023) concurs, creating good prompts is not easy, with some artists viewing them as artistic products.

One of the benefits of such generative AI systems, however, as Tosca (2023) highlights, is that they '*can provide the element of randomness that is necessary to take us out of our usual mental structures and augment creative output*' (p. 163). As the CEO of Midjourney, David Holz has said, Midjourney '*is designed to unlock the creativity of ordinary people by giving them tools to make beautiful pictures just by describing them*', which serves the mission of '*trying to expand the imaginative powers of the human species*' (quoted in Salkowitz, 2022). So, while prompts can produce more or less detailed, or original results, like Tosca, Tigre Moura (2023) has proposed that, in view of their ease of use, AI tools motivate 'non-creatives' to express intrinsic intents through artificial co-creation.

Concerns about ML Art

While the ease of using ML platforms may be motivating for 'non-creatives', West and Burbano (2020) have questioned whether the ease of using such technologies will lead to

> *lack of awareness of potentially encoded bias or decreased artistic agency in accepting whatever aesthetics are 'baked' into the ML algorithm by its designers? Will this lead us to unconscious application of a homogenized ML aesthetic as it is sublimated into digital content creation tools?*
>
> (p. 3)

These questions are particularly pressing for text-to-image models because, as Heaven (2022) explains, most of these models today

> *are trained on a large data set called LAION, which contains billions of pairings of text and images scraped from the internet. This means that the images you get from a text-to-image model are a distillation of the world as it's represented online, distorted by prejudice (and pornography).*

This has certainly been reflected in the experiences of users of this technology, who have reported biases showing up in the results. For example, writer Sarah Cascone (2022) found that Lensa – an app that runs on Stable Diffusion, which was trained on LAION 5B, to produce avatars in different styles – tended to reflect certain societal biases by producing objectifying images of women and racial caricatures. As well as ethical issues posed by using such datasets, other commentators have elaborated on the aforementioned aesthetic issues.

While both humans and generative AI systems can represent non-existent and existing objects and scenes alike, Manovich (2023) has highlighted that rarities in the training data of the latter are hardly learned or not even parsed at all, and so cannot be synthesized. In fact, the results of text-to-image models can be quite repetitive, and this may, as Maria Teresa Llano, a student of computational creativity says, in Heaven (2022),

> *be a fundamental limitation of the technology. By design, a text-to-image model churns out new images in the likeness of billions of images that already exist. Perhaps machine learning will only ever produce images that imitate what it's been exposed to in the past.*

Arielli (2021) has explained that ML systems are ideally suited to produce traditional artworks as they tend to display a good amount of repetition and similarity. Relatedly, as curator Luba Elliott has pointed out of the researchers who develop image generation models: '*They focus on the aesthetics of an image. They're much more interested in trying to replicate the styles of paintings of the past than in participating in current developments in contemporary art*' (Ploin et al., 2022, p. 11).

It is worth highlighting here, however, that there are plenty of contemporary artists who are primarily image-makers, working at the cutting edge of current developments in contemporary art and engaging with the history of images in doing so. Moreover, as Langland-Hassan (2020) points out, '*most art – even great art –* is *highly imitative in nature [. . .] An artist's work is understood, in the first instance, in relation to her influences*' (p. 290). Indeed, rather than appearing from cultural vacuums, '*all art depends upon the learning of prototypes*' (Kupferberg, 2021, p. 103). Importantly, however, the process of imitation, as Fronda has outlined, is more than just copying: '*An imitator was expected to emulate many models, join imitated material seamlessly to his own, reshape and vary it for its new context, and improve upon it*' (quoted in Tosca, 2023, p. 142). This process enables one to gain an understanding of how a genre works and so to later be able to develop one's own style within that genre (Tosca, 2023).

There are perhaps two senses in which prototypes are involved in artificial image generation: (1) there are prototypes for the processes of generating the images, be that through prompts or any of the five activities outlined earlier – crucially, as Manovich (2022) highlights, digital media externalizes the thoughts and processes of an individual so that their actions can be studied and applied – and (2) there are also the images in the training data, which are emulated in the generated images. The artists working directly with GANs and those writing prompts can emulate their models, as per (1) in the process of generating images (e.g. by studying another's code or method of crafting a prompt), however, this changes for (2) where frequently the processes will differ. For example, a user may aim to emulate a certain style of painting and write a prompt accordingly to generate an image, however they are not undertaking the same kinds of image-making processes (i.e., applying paint to a support), which potentially limits their ability to gain an understanding of how that practice works.

Moreover, for those using text-to-image platforms, the contents of the dataset that informs the output is not chosen, or even potentially knowable, by the user. This is different for the artists working directly with GANs, who not only actively choose their datasets, but in doing so, aim to capitalize on, and interrogate, the features of the technology itself and not merely, as users of text-to-image models might legitimately aim to produce, a picture that matches the text description. Significantly, as highlighted in the cases of Anna Ridler and Jake Elwes earlier, they determined what kinds of images would be generated (without specifying the exact features that would result) by choosing to constrain the scope of the dataset. In fact, the ability to choose such constraints seems key to using these systems as creative aids and complements the proposal that the achievement of novelty in art can be understood in terms of imaginative constraints, whereby the right balance is achieved between releasing and imposing constraints (Kind, 2022).

As Kind (2022) highlights, while creativity in the domain of art may seem to rely on transcendent uses of imagination, different artistic ends call for different degrees of constraints:

> *While some art strives for realism, some does not. An artist working on a series of realistic paintings set in New York will thus need to constrain their imaginings in a way that an artist working on a series of fantastical paintings set in the lost city of Atlantis will not.*
>
> (2022, p. 16)

Analogously, Anna Ridler constrained the dataset to 10,000 photographs she took of tulips to create the work *Mosaic Virus* between 2018 and 2019 and produce an evolving set of images of plausibly real tulips, while David Young trained his system with no more than a handful of images, consisting of solid colours or basic shapes, to create the abstract series *Tabula Rasa*.

Creativity and Constraint

More broadly, McLeish (2019) has argued that no art results from unconstrained exercises of imagination and relatedly, the view that constraints can enhance artistic creativity, has received increasing support from theorists across a range of disciplines (Elster, 2000; Stokes, 2007; Livingston, 2009; Boden, 2010; McLeish, 2019; Kupferberg, 2021). Constraints, it has been proposed, can inspire and orient endeavours by helping artists to avoid tired personal mannerisms or repetitive stylistic habits (Livingston, 2009) and establishing '*formal as well as substantive, or content-related parameters, and corresponding normative expectations and dispositions*' (p. 138). Crucially, the latter can help to prevent '*the inability to experience a spontaneous judgment of the results of artistic experimentation*' (Livingston, 2009, p. 140) by attuning and enhancing activation '*of a capacity for response not directly under the artist's conscious control*' (Livingston, 2009, p. 141). Livingston has noted that not all constraints need be arrived at independently: '*Some such constraints are inherent in a specific technology or craft; others may be proposed or imposed by a patron or producer, while others are more independently arrived at by the individual artist or group of collaborating artists*' (2009, p. 137).

Elster has formalized this idea through a tripartite approach distinguishing between constraints that are self-imposed, imposed from outside, or intrinsic (like Livingston's inherent constraints), and proposed '*that both choice of constraints and choice within constraints*' is how artists try to maximize artistic value (2000, p. 178). In addition to the constraints already mentioned, these may include hard constraints, like '*formal, material, technical, or financial restrictions on the selection and combination of the constituent units of the given medium*', or soft constraints, such as conventions or '*restrictions that constitute a specific genre*' (Elster, 2000, p. 190). Notably, although Elster discusses artistic value in terms of aesthetic value, which he conceives of as the production of emotional effect (2000, p. 205), his account could also work with a pluralistic conception of artistic value, where several values, like aesthetic value, cognitive value, and art-historical value, for example, might constitute the artistic value of a work (Stecker, 2019) – a conception that seems to be more commonly held in the contemporary art world.

Based on the foregoing, artists working directly with GANs have plenty of opportunities to select and balance their constraints – including the models they select or build and the kinds of images that will be produced based on the choice of dataset – and make decisions within the constraints that come with these technologies – from the size of the datasets used or constructed to the length of time the models are trained, for example. There is evidence that those working with text-to-image models can also choose their constraints, by deciding to work with DALL·E rather than Midjourney, for example, and make choices within the constraints of these models. For instance, Eldagsen, a photographer by trade, has used his practical and linguistic knowledge of the domain to skilfully shape the way AI image generators, trained on billions of pairings of text and images, would 'interpret' his prompts. This is perhaps key to understanding how he was able to produce such '*immensely evocative*' (Williams, 2023) images. Eldagsen deployed his knowledge and skills to work within the constraints imposed by the prescribed datasets and produce an innovative body of generated images that drew on the visual lexicon of photography. So, what can be taken from this discussion to progress the use and development of generative AI models as imaginative and creative aids to fulfil the 'democratizing' (Pilat, 2022; Ploin et al., 2022; Tigre Moura, 2023, p. 1; Khosrowi et al., 2023) potential of this image-making technology?

Future Research Agenda

Readers may be quick to point out that there is a difference between the use of these technologies by artists and 'non-creatives'. Certainly, systems like Midjourney can aid users when they are unable to imagine how an image corresponding to a prompt could look. In such cases, what the technology is doing is helping a user to *visualize* their prompt. However, while commonly used as synonyms, imagination need not implicate visual imagery (Gaut, 2003). There are, therefore, two distinguishable senses in which generative AI systems can be used as imaginative aids:

(1) To aid a user in visualizing an idea
(2) To aid a user's creative process by assisting them in cognitive play

The latter may involve the former, but the former need not entail the latter. Point (1) is something that text-to-image platforms can potentially do very well. There is also some scope for these platforms to do. With regard to (2), for example, if a result comes back that does not align with a user's ideas, it may help the viewer to isolate potentially valuable features and imagine a composition incorporating more of those features or produce alternative imaginative associations, and hence assist them in the cognitive manipulation that, as Stokes argues (2014), is vital for creativity.

Artists working directly with GANs can certainly do (1) (although perhaps less reliably than text-to-image models) and have far greater potential for using these systems as per (2), given the wider range of choices that facilitate artists in balancing constraints. Furthermore, as Elgammal (2023) and others have highlighted, these models are more likely to return surprising results – often due to mislearning and failures – that can really spur on cognitive play.

Readers may question whether (2) is really as important for 'non-creatives' or perhaps casual users who would simply like to produce some pleasing images without having to pick up a pencil or camera. There is a case to be made that the development and facilitation of the use of these systems ought to make (2) more achievable for all users, as doing so likely entails that they are not only enhancing their capacity to be creative in other domains, but also that users will have had to engage more deeply in the practices they are emulating and the algorithmic worlds they are engaged with, not only through these platforms, but in everyday life also, as the following suggested research agenda demonstrates.

RQ1: What can we learn from artists who are creatively working with ML? As the preceding discussion demonstrated, artists can skilfully navigate these new technologies to identify points which offer scope for autonomous choices of, and within, constraints, many of which are inherent to this technology, or imposed by prescribed Internet-based datasets. There is, accordingly, much that can be learnt from their skills and knowledge that allow them to work so creatively with these models. Many pioneering artists in the field of ML art have highlighted the benefits that would come from more interactions between artists and engineers. As indicated earlier, and as Jake Elwes has underscored in Ploin et al. (2022), many engineers use historic standards to validate this new medium, much like photographers initially tried to mimic paintings. In addition to offering more contemporary standards and concerns for this developing practice, artists can also offer alternative perspectives on these potentially disruptive technologies (Ploin et al., 2022). One barrier that is often mentioned is lack of a common language (Ploin et al., 2022). Sometimes, however, these barriers have been overcome, as in the case of the development of Artrendex's accessible generative AI platform Playform. As Elgammal and Mazzone (2020) outline, they developed this platform to facilitate the work of contemporary visual artists and, consequently, it was designed to offer a broader range of choices, so that artists could generate quickly and with fewer images, either from databases of images already built into the system or with their own image databases, using an intuitive interface. This, Elgammal and Mazzone (2020) have convincingly demonstrated, facilitates more artistically relevant choices and, by extension, productive feedback loops between the human and generative system. Accordingly, more attention to such cases would inevitably aid research that addresses how to bridge that gap and foster more fruitful collaborations.

RQ2: How will the popularity of text-to-image models affect developments in ML art?

While for many it is liberating to be able to produce images from text, this does not necessarily serve everyone, or even art, well in the long run. Elgammal (2023) has highlighted that artists tend to be visual thinkers and has argued that '*A.I. now creates its visual output confined by our language, losing its freedom to visually manipulate pixels freely without prevarication from human semantics*'. As the foregoing has shown, while constraints may be beneficial for artistic creativity, the opportunity to choose those constraints, such as whether to be confined to language in the process of generating images, can help individuals to channel their imaginations in productive new directions. Accordingly, as Manovich (2023) suggests, it would be fruitful to continue developing AI for mappings between all types of media, without privileging text, to stimulate imaginations.

Another issue relating to text-to-image models is the enormous datasets they are trained on and the extent to which users are aware of what these consist of. One of the key differences between artists working directly with GANs and users of text-to-image platforms is the ability to determine what goes in the dataset. There is certainly scope for users to fine-tune systems like Stable Diffusion, so that the biases and noise of such massive networks 'do not dominate our own imagination' (Manovich, 2023, p. 17). However, at present fine-tuning these systems with one's own data is not easy and so further research into this would be a huge step forward in enabling users to gain more control over how they use these systems. There is undoubtedly value in repetition and sameness, as Tosca (2023) has convincingly argued; however, text-to-image platforms would ultimately be more democratizing, as a means of producing images of existent or non-existent objects with certain kinds of features, if users are enabled to make some of the key decisions, such as what data are used to inform the model, that would allow them to develop and pursue their own creative aims.

RQ3: How do AI image generators affect how users represent their own agency?

Due to the sophistication of text-to-image platforms, impressively realistic or stylistic images can be generated from a simple prompt. So, without relevant counterfactuals, such as how the images would have looked if a similar prompt had been entered or information about the training data, Khosrowi et al. (2023) have proposed that such users may suffer from illusions of creatorship when they have made minimal contributions to an output. Relatedly, for the kinds of users just outlined, attention should be paid to how these platforms may facilitate unhelpful attitudinal changes to image and art-making. Considering that some of the platforms are marketed or widely discussed as creating art, some users may consider what they have done a successful artistic achievement if they manage to generate an image that matches the basic description they supplied. However, as we have seen, the most innovative results require skills and knowledge of these domains. By reducing these skills, and the rich and subtle values that are involved in creative image and art-making, these activities may be said to possess value clarity (Nguyen, 2020), which entails that the application of goals is obvious (generate images that match the description in the prompt), easily commensurable (did the features of the image match the description?), and rankable (did some images match the description more closely?).

Value clarity, as Nguyen (2020) cautions, is useful when possessed by a game, but outside of a game, however, it '*may infect choices we make as to which enduring values and goals we should take on*', which can in turn, '*quietly undermine our aims and our autonomy*' (pp. 199–200). This issue is particularly pertinent given that for many, without even realizing it, ML is inbuilt into their daily interactions with image-making technologies, such as Photoshop, Blender, and smartphone photography (Manovich, 2022). Consideration should thus be given for how to frame these activities to ensure that individuals develop realistic attitudes towards their contribution to the generation of images. Greater transparency and improved representations of the scope of image generating may help here – as Ploin et al. found: '*perception of agency on the part of the model diminished with artists' technical understanding*' (2022, p. 49). Active nudges on platforms could be helpful in motivating users to pursue alternative workflows and investigate the practices they are emulating in their activities. As the foregoing has underscored, rather than replacing traditional image-making, some of the most creative work is happening when skills and knowledge relating to these processes inform AI image generation – future research should seek to facilitate more interconnections between these areas for users.

Acknowledgements

The author is thankful to the editor of this volume, Francisco Tigre Moura, and also to the audience at the University of Oxford Aesthetics Seminar (2023) for their helpful comments on an earlier version of this chapter.

References

Anscomb, C. (2021). Creative agency as executive agency: Grounding the artistic significance of automatic images. *The Journal of Aesthetics and Art Criticism*, *79*(4), 415–427. https://doi.org/10.1093/jaac/kpab054

Arielli, E. (2021). Extended aesthetics: Art and artificial intelligence. *Proceedings of the European Society for Aesthetics*, *13*, 1–13.

Boden, M. (2004). *The creative mind: Myths and mechanisms*. Routledge.

Boden, M. (2010). *Creativity and art: Three roads to surprise*. Oxford University Press.

Cascone, S. (2022, December 8). I uploaded photos of myself to the New Lensa A.I. portrait generator: The results were stunning, strange . . . and super creepy. *Artnet News*. https://news.artnet.com/art-world/lensa-ai-avatar-results-2225393

Coeckelbergh, M. (2023). The work of art in the age of AI image generation: Aesthetics and human-technology relations as process and performance. *Journal of Human Technology Relations*, *1*(1), 1–13. https://doi.org/10.59490/jhtr.2023.1.7025

Elgammal, A. (2023, June 20). Text-to-image generators have altered the digital art landscape – but killed creativity: Here's why an era of A.I. art is over. *Artnet News*. https://news.artnet.com/opinion/ahmed-elgammal-op-ed-ai-art-is-over-2304028

Elgammal, A., & Mazzone, M. (2020). Artists, artificial intelligence and machine based creativity in playform. *Artnodes*, *26*, 1–8. http://doi.org/10.7238/a.v0i26.3366

Elster, J. (2000). *Ulyssess unbound: Studies in rationality, precommitment, and constraints*. Cambridge University Press.

Gaut, B. (2003). Creativity and imagination. In B. Gaut & P. Livingston (Eds.), *The creation of art: New essays in philosophical aesthetics* (pp. 148–173) Cambridge University Press.

Heaven, W. D. (2022, December 16). Generative AI is changing everything: But what's left when the hype is gone? *MIT Technology Review*. www.technologyreview.com/2022/12/16/1065005/generative-ai-revolution-art/

Khosrowi, D., Finn, F., & Clark, E. (2023). Diffusing the creator: Attributing credit for generative AI outputs. *Proceedings of the 2023 AAAI/ACM Conference on AI, Ethics, and Society*, 890–900. https://doi.org/10.1145/3600211.3604716

Kind, A. (2022). *Imagination and creative thinking*. Cambridge University Press.

Kupferberg, F. (2021). *Constraints and creativity*. Cambridge University Press.

Langland-Hassan, P. (2020). *Explaining imagination*. Oxford University Press.

Livingston, P. (2009). Poincaré's "Delicate Sieve": On creativity and constraints in the arts. In M. Krausz, D. Dutton, & K. Bardsley (Eds.), *The idea of creativity; philosophy of history and culture* (pp. 129–146) Brill.

Manovich, L. (2022). Who is an artist in software era? In E. Arielli & L. Manovich (Eds.), *Artificial aesthetics: A critical guide to AI, media and design*. http://manovich.net/index.php/projects/artificial-aesthetics

Manovich, L. (2023). AI images and generative media. In E. Arielli & L. Manovich (Eds.), *Artificial aesthetics: A critical guide to AI, media and design*. http://manovich.net/index.php/projects/artificial-aesthetics

McLeish, T. (2019). *The poetry and music of science*. Oxford University Press.

Nguyen, C. T. (2020). *Games: Agency as art*. Oxford University Press.

Pilat, A. (2022, October 27). I make art with A.I. here's why all artists need to stop worrying and embrace the technology. *Artnet News*. https://news.artnet.com/news/artists-embrace-ai-dall-e-op-ed-2199088-2199088#:~:text=Like%20any%20new%20technology%2C%20A.I.,unique%20tool%20to%20shape%20culture

Ploin, A., Eynon, R., Hjorth, I., & Osborne, M. A. (2022). *AI and the arts: How machine learning is changing artistic work*. Report from the Creative Algorithmic Intelligence Research Project. Oxford Internet Institute, University of Oxford.

Salkowitz, R. (2022, September 16). Midjourney founder David Holz on the impact of AI on art, imagination and the creative economy. *Forbes*. www.forbes.com/sites/robsalkowitz/2022/09/16/midjourney-founder-david-holz-on-the-impact-of-ai-on-art-imagination-and-the-creative-economy/

Stecker, R. (2019). *Intersections of value: Art, nature, and the everyday*. Oxford University Press.

Stokes, P. D. (2007). Using constraints to generate and sustain novelty. *Psychology of Aesthetics, Creativity, and the Arts*, *1*(2), 107–113. https://doi.org/10.1037/1931-3896.1.2.107

Stokes, P. D. (2014). The role of imagination in creativity. In E. Paul & S. Kaufman (Eds.), *The philosophy of creativity* (pp. 157–184). Oxford University Press.

Tigre Moura, F. (2023). Artificial intelligence, creativity, and intentionality: The need for a paradigm shift. *Journal of Creative Behaviour*, *57*, 336–338. https://doi.org/10.1002/jocb.585

Tosca, S. (2023). *Sameness and repetition in contemporary media culture*. Emerald Group Publishing.

Williams, Z. (2023). "AI isn't a threat" – Boris Eldagsen, whose fake photo duped the Sony judges, hits back. *The Guardian*. www.theguardian.com/artanddesign/2023/apr/18/ai-threat-boris-eldagsen-fake-photo-duped-sony-judges-hits-back

Chapter 7

AI and Art

Non-Human, Still Human, or All Too Human?

Caterina Moruzzi and Oreste Leone Campagner

Introduction

Since the beginning of the twenty-first century, the accelerated advancements in artificial intelligence (AI) technology, coupled with its increased use in the creative industry, have caused a shift in the way art is produced and consumed. This chapter reflects on how AI models for content generation are transforming traditional approaches to artistic and creative production, inevitably affecting the role of artists and the nature and function of art (Boutin et al., 2023).

After the company OpenAI presented its text-to-image generative system, DALL·E 2, in April 2022, other major tech companies rushed to release their own versions: Google Imagen, Midjourney, and Stable Diffusion are just a few examples of the generative AI models which have been downloaded and used by an impressive number of professionals, amateurs, and enthusiasts in the past year. The user-friendly nature of these platforms greatly contributes to the fast rate of their adoption. Generative AI models have been leveraged by artists in their practices for much longer (Franco, 2022). However, it is only with text-to-image systems, which convert in only a few seconds users' text prompts into images matching the given description, that a wide and not necessarily computer-literate audience has been given easy access to this technology.

The "generative revolution" phenomenon has attracted substantial media coverage, with numerous blog posts, podcasts, and newspaper articles daily being released on the subject. However, in many instances, these pieces tend to add to the hype around generative AI instead of truly informing their readers. This chapter aims to provide an alternative to hyperbolic proclamations through a more nuanced consideration of the long-term impact of this technology on the nature and function of art and artists.

The chapter starts by considering three options for interpreting how artists and art made with AI may be positioned in the art world after the widespread adoption of AI in various creation processes (Section "Human or Non-Human? Exploring a Third Option"). It then goes on by considering a

DOI: 10.4324/9781003453901-10

paradox intrinsic in the last of these three options: art after AI seems to be less human, but it can be argued that it is instead more human than before (Section "Solving an Apparent Paradox"). The chapter closes by observing how some of the traditional functions of art – communication, knowledge, and transcendence – despite manifesting themselves differently, are formally preserved in this landscape, if not enhanced (Section "The Function of Art after AI").

Throughout the chapter, the term "AI" will be used for brevity, and it will refer to a specific application of AI technologies, namely text-to-image systems, such as diffusion (Rombach et al., 2022) and transformer models (Radford, 2021) for content generation. In addition, the chapter will intentionally develop the argument in relation to "art" rather than "creativity", as it concerns the transformation brought by the development of AI technology to the nature and function of art, although it is worth noting that generative AI models are used by creatives in many fields beyond the artistic one. This means that our conclusions do not necessarily apply to the broader notion of creativity.

Art and Artists after AI

From photography to portable paint tubes, and up to graphics editor software, artists have leveraged technological innovations as tools to express their artistic concepts (Candy & Edmonds, 2002). The connection between the fields of art and technology is more profound than what might appear at first. Indeed, the etymology of the term "technology" traces its roots back to the Greek word *techné*, which, in ancient Greece, was also used to refer to the activities of painters and sculptors, who eventually became representatives of the arts in modern times (Tatarkiewicz, 2012).

The recent emergence of AI models for content generation might be interpreted within a linear evolution of the use of technology in the arts. However, the remarkable speed at which these models are being adopted and employed leads us to ask whether the consideration of art as an exclusively human endeavour should be revised. To address this issue, we need to account for the transformation that not only art but also the figure of the artist might be subject to. Indeed, in Gombrich's words (1995), the history of art is also the history of artists. In the next section, the chapter thus explores three options in respect to the nature and function of art and to the figure of the artist in the landscape that is shaped by the broad use of AI models for content generation: (1) non-human, (2) still human, and (3) Human^n.

Human or Non-Human? Exploring a Third Option

The first option, Non-human, depicts AI as the sole responsible for the artistic process and for the output that results from it, while the role of humans in the artistic process is significantly decreased in scope and significance. This option is admittedly also the most controversial of the three and, to the authors' knowledge, nobody has put forward thoughtful arguments in support of this claim. Yet it is unquestionable that the many sensationalist headlines predicting a bleak future for human artists replaced by their AI counterparts greatly contributed, at least among the general audience, to instilling the fear that a future without human artists is conceivable, if not probable. Still, a supporter of this first option can find some arguments in its favour.

A laudable aspect of acknowledging the possibility for AI to be deemed the sole artist is its evasion from a traditional anthropocentric vision. In the literature on creativity, many are the voices sympathetic to attributing creative skills to non-human animals (like corvids and octopi), and, potentially, also to artificial entities (Kaufman & Kaufman, 2015; Simon, 1985). And yet the resistance against ascribing *artistic* skills to non-human agents is greater. However, one could still advocate for the attribution of the title of "artist" to the AI by hinting at the growing autonomy of recent models in elaborating the data provided by humans during the training phase, resulting in the creation of artefacts that are shared, discussed, and even awarded, just as human artworks are. In this view, roles are reversed: the human is the tool leveraged by the AI, insofar as humans give the AI the instruments to work with, that is the annotated dataset and the computational structure (Wan et al., 2022).

Unlike the first option, Non-human, the second option has gained widespread consensus among artists and researchers alike: art is still human. While AI can offer new and exciting possibilities, it is ultimately the human artist who creates the work. According to this view, AI is not different from previous technological tools that enabled various art forms, like etching or photography, or that opened new paths for already-established ones, like synthetic paints or photo editing software (Hertzmann, 2018).

These first two options lead to opposite conclusions: either the human or the AI is recognised as the artist. However, it can be argued that both of these options fail to account for the complexity of the phenomenon of the increasing adoption of AI in artistic processes. This chapter therefore proposes a third, more nuanced option that tries to capture both the essentially human nature of art and, at the same time, the growing role of technology in creating it. Art is Human^{n}.

Some arguments depict the relationship between humans and AI models in creative endeavours as a collaboration where AI might be deemed a co-creator but not the primary one (Anscomb, 2022; Mazzone & Elgammal, 2019). Here something different is argued. It is not a matter of priority: in this relation AI is indeed external to the human artist, but it contributes to the artistic process insofar as it is the expression of a collectiveness of human ideas, products, and experiences, which exponentially amplifies the humanity which art encapsulates and exhibits. This is what the "n" exponent in the

Human^n attempts to represent. Technology has an increasing influence on creative processes, but its role does not diminish the relevance of the human contribution, rather it enhances it.

In the creation of content with AI, the agency of the creative process is shared between humans and technology at an unprecedented level. For this reason, the latter cannot be considered a mere tool, as option (2) detailed above entails. On the other hand, the human counterpart still retains a fundamental role in the artistic process and, therefore, the label of "artist" cannot be solely attributed to the AI that outputs the final product of this process, as option (1) provocatively suggests.

The proposal of a third option can be further elaborated, in the attempt of clearing up a paradox that seems to emerge: with the use of generative AI models, art might feel *less* human than before. Thus, how can it be argued that, on the contrary, it is *more* human than ever, and specifically Human^n?

Solving an Apparent Paradox

The enhanced humanness of art made with AI can be observed both in the creation and in the reception of it. Starting with the latter, AI-generated content forces the human observer to play a more active role in interpreting and curating it than ever before. The relevance of the beholder in art is well established in the literature (Zangwill, 1999). However, it might be argued that it is even more crucial now, as the mass production of AI-generated content challenges the observers' ability to frame it within an artistic and cultural context. Algorithmic platforms offer the audience the possibility of becoming part of the art world through their interaction with the pieces shared on the web (Becker, 2008). The audience does not simply decide what is art but is ultimately the one who makes the artwork by experiencing, liking, and commenting on it.

The powered humanness of art made with AI can also be observed in terms of content creation: text-to-*x* (image, sound, speech, etc.) models exploit a training set which includes an immense collection of multimodal information stemming from human experiences. For example, LAION 5B, the large-scale dataset that many of the text-to-image models are trained on, consists of 5,85B image-text pairs. Scholars have already used the metaphor of the mirror to contend how the behaviour and idiosyncrasies of humans are reflected by AI (Vallor, 2022). However, aspects that are not immediately visible or even accessible may be exposed (see Section "Transcendence"). AI seems to be a peculiar kind of mirror, similar to the Mirror of Erised, which, in the *Harry Potter* saga, does not render a faithful image of what stands in front of it but rather shows what the person wants to see ("erised" is "desire" spelled backwards) – or maybe, in the present case, what the algorithm *lets* them see.

The third option for interpreting the process of creating art with AI (Human^n) may therefore help to capture the complex dynamics behind the recognition of the active influence of AI in the artistic process, without downplaying the human contribution. In fact, it offers a way to address the concern raised by Sarkar (2023), among others, regarding the danger of overlooking the labour performed by human annotators when fuelling the narrative of AI as a "collaborator". The mentioned paradox is only apparent, as the art resulting from the human–AI collaboration is more human rather than less. AI is ultimately a human creation and, as such, it can be seen as an extension of humanity, rather than a completely alien being. AI models may indeed be deemed co-creators, but they are co-creators which are created through a multitude of human experiences.

The Function of Art after AI

The consideration of the essential attributes of art is inextricably tied to the examination of which function it plays in society. This section focuses on three of the many domains in relation to which art has traditionally played a critical function, analysing which changes, if any, such functions are subject to under the influence of AI technology: (1) communication, (2) knowledge, and (3) transcendence. The reflection on these functions and their transformation will reinforce our proposal that art is moving toward a $Human^n$ dimension.

Communication

One of the key ways in which art fulfils a communicative role is in providing a platform for artists to express their intentions and to share them with an audience. For the message to reach the public, however, artists need to establish an exchange also with the medium that makes the communication possible. In addition, typically the audience does not passively receive the artists' message but is instead actively involved in an act of interpretation (see section "Solving an Apparent Paradox"). The work of art itself is a speaking entity in which the observer interprets a meaning. This meaning can be literal – as it was for religious art, especially in a past where religiously themed artworks had the task of narrating the sacred texts to a mostly illiterate public (Campbell, 2001) – but it can also be symbolic, as it is more common since the advent of non-representational art movements. Even after the metamorphosis experienced through technological, social, and cultural transformations, art still maintains a communication function which articulates in the three fundamental components of creator (or sender), medium, and audience (or receiver). However, while preserved, these dimensions have changed.

The practice of prompting can be used to illustrate this transformation. Typing a text prompt through the interface of the model induces the latter to generate content, be it images, videos, text, or music, matching the instructions given as accurately as possible. The prevalence of this modality of interaction between the user and the AI model led to the recent emergence of prompt-engineering, which is now already a well-established practice with a dedicated community, textbooks, courses, and conferences (Oppenlaender, 2023).

In this practice, both the modality of transmission and the control the creator has over it are different in respect to previous methods used by artists to employ technology in their artistic processes. Indeed, the primary modality users have to interact with the new generation of generative AI models through natural language, while other art forms for the production of content, such as painting, photography, sculpture, video, or printmaking, may require the use of natural language only in some phases of their processes.

In addition, due to the nature of the interaction, restricted to giving text instructions, the prompter's control over the final output is limited. In order to write effective prompts, users need to not only have knowledge of the training set and of the model's underlying architecture but also master the vocabulary of prompt modifiers, that is keywords or key phrases that are not evidently linked to the appearance of what the prompter wants to achieve but have the effect of increasing the prompter's control on the desired features of the output. The lack of fine-grained control by prompters on the final product is highlighted by the copyright registration guidance policy issued by the US Copyright Office in March 2023 to clarify practices for registering works that contain material generated by the use of AI technology: "[P]rompts function more like instructions to a commissioned artist – they identify what the prompter wishes to have depicted, but the machine determines how those instructions are implemented in its output" (Copyright Registration Guidance, 2023). For the communication between prompter and program to be effective, and for prompters to keep as much charge as possible of the features of the output, they need to master the language of AI (El-Assady & Moruzzi, 2022). This engineering aspect of the practice brings prompting artists closer to the ancient blending of art and craft embedded in the concept of *techné*.

With regard to the audience, as discussed, it necessarily needs to adopt an ever more active critical attitude in interpreting the plethora of AI-generated content it is flooded with. The audience's critical sense needs to be switched on in the effort of not only interpreting what the artists want to communicate through the artefact they create, or which meaning the latter might have in today's society, but it also becomes essential to decipher the provenance and authenticity of artistic content available online. Succeeding in interpreting the artists' message and decoding the provenance and authenticity of a piece is made more complex by the limitation of the control that a creator can have on the machine and by the opacity of the AI models' architecture. The effort the audience needs to employ in this exegetic endeavour makes the relevance of the communication aspect of art even more prominent.

Knowledge

Art has historically carried out an epistemic function, the aesthetic experience being associated with knowledge production and with the engagement of our understanding (Perkins, 1988). Before modern times, art mainly attempted to represent the outer world, that is an other-than-self, be it a physical object, animal, landscape, a historical or religious scene, or to search for a principle of harmony and perfection in what exists (Tatarkiewicz, 2012).

The switch to non-representational art forms implied a shift in focus from the outer to the inner world. Since the eighteenth century, and with Romanticism, feelings and the self started to become the work of art (Kuperus, 2016). Following Freud's theories and the crisis of the individual at the beginning of the twentieth century, the psychological element was evident in some of the historical vanguards, such as Expressionism. Not only feelings but a broad range of mental aspects are involved, from the oneiric settings of Surrealism to Conceptual Art, where the aesthetic experience is fully non-representational and focuses purely on thought and the understanding of the self.

The introduction of AI in artistic processes seems to bring together both historical moments of the epistemic function. On the one hand, to find the best prompt, humans need to clarify what they want to be seen in the output. Prompting therefore fulfils a self-reflective function, linked to its use of natural language (see Section "Communication"). The role of speech in thought creation is effectively summarised by the German writer Heinrich von Kleist, parodying the French *l'appétit vient en mangeant* ("appetite comes with eating") into *l'idée vient en parlant* ("the idea comes with speaking") (von Kleist & Hamburger, 1951). By interacting with the machine, humans eventually understand themselves better.

On the other hand, knowledge about the external world corresponds now to the inquiry into AI itself. Art is the mediating element between humans and machines, the process through which the functioning schemes of algorithms are investigated and made more transparent (Zylinska, 2020). While finding the best prompt, the community of creators is generating knowledge on an other-than-self (the AI) *through* the creative process and, ultimately, art.

However, recalling the paradox mentioned in Section "Solving an Apparent Paradox", it is evident that this other-than-self represented by AI is nothing else than the mirror of humanity that has been discussed. Consistently with the Human^n category, self-reflection occurs in prompting as a twofold process, creating knowledge both about human intentions and about the AI, that is to say, once again, about humans. The traditional communicative function of art is therefore not only preserved but even enhanced.

Transcendence

Besides the communicative and epistemic, this section considers the transcendental function. The term "transcendental" is not used here in the Aristotelian connotation of "properties of being" but rather in the more common-sense meaning of goal, aim, or purpose. Under this function, a variety of manifestations of the aesthetic experience is frequently connected to an existential or mystical dimension. Aesthetics also consist of experiences that cannot fully be articulated in language and are therefore perceived as ineffable

(Gunther, 2017). Such inexplicability is reminiscent of the three transcendental moments identified by Thomas Rentsch with reference to negative theology, namely the transcendence of object, language, and sense (Hailer, 2014). Holding to a purely philosophical and non-confessional perspective, this is arguably often encountered in the aesthetic experience: *something* happens that cannot be fully understood or explained; underlying the experience there *is* a sense which cannot be entirely grasped. Be it historically interpreted as the contact with a divine, sublime, unconscious, or "ultimate truths", the transcendent function of art gives access to a "special" dimension.

Instances of contents made with AI may still be conducive to such an access. A first level of transcendence occurs in the very interaction with the machine which, as argued earlier, is ultimately a projection of humanity itself: humans are transcending themselves in the literal sense of going *beyond* their bodies and externalising onto a collective self where they can reflect themselves. The second level of transcendence occurs when AI reacts to the input. Due to the opacity of its internal processes, however, the result is hardly ever predictable. This is where the unspeakable happens, and the user is left puzzled: faced with many of the outputs of generative AI, they are accessing *something* that they cannot fully describe in words.

A practical example may help to clarify. In 2022, while working with a generative AI model, the Sweden-based artist, Steph Swanson, known as Supercomposite, reported continual appearances of the same figure, now referred to as "Loab". This was showing up as a sad, sobbing-looking woman and was recurring in further prompt cycles, despite Swanson's attempts to dilute her presence in the algorithms. More shockingly, she was often depicted in horror-like scenes, featuring dead or injured children, which contributed to extended coverage in the media (Lavoipierre, 2022). What is so unsettling about Loab? Is it the access to the most morbid parts of the unconscious, or the fact that such shocking images were realised by a non-thinking external entity? The transcendent moment is well explained here by Swanson herself, when she claims that "it's almost scarier than being supernatural in some way. It is truly unknowable" (Lavoipierre, 2022).

This ineffability might be related to the Uncanny Valley, a concept developed to explain one's sudden revulsion to human-looking robots (Mori et al., 2012). The uncanny originates here from the Human^n paradox: in the dialogue with the machine, a tension originates between one's tendency to anthropomorphise AI and the realisation that this is a human artefact; concerning the outputs, one is confronted with images traditionally considered as human-made, whereas now they are not first-hand created by humans in the strict sense. The uncanny is a feeling of uncertainty about what it is to be human: it is a "crisis of the proper" (Royle, 2003) assuming that making art is deemed an inherent property of humanity.

The uncanny also results from made-with-AI artistic content commonly exhibiting glitches and mistakes, for example when depicting hands.

Mistakes are a fundamentally human trait which generates the unexpected in the creative process (vanEenoo, 2012). By identifying errors in AI outputs, the human element which allows humans to self-reflect, understand, and transcend themselves is captured. Through AI mistakes, humans are offered the gateway to the unspeakable. Additionally, errors can boost the creativity of human artists, as testified by artistic practices based on glitches like the works of Dom Barra or exhibitions like Unit London's *The Perfect Error* or by the debate, internal to prompters' communities, about which prompt is needed to replicate or exclude a given mistake, as was the case with Loab (Oppenlaender, 2023).

To conclude, the three traditional functions of art – communication, knowledge, and transcendence – are not only preserved despite changes but also are strongly enhanced in the landscape of art generated with AI, in virtue of its being Human^n. It is worth making a final consideration here. Communication and knowledge are functions associated with a plethora of other human activities: for instance, a cookbook. What distinguishes art is namely transcendence, which can be accessed in the case of generative media, as discussed, through the occurrence of glitches. With the refinement of generative AI models, however, errors would become increasingly rare: the risk is therefore for the use of AI in artistic practices to lose its creativity-boosting feature, as well as a consistent part of its transcendental function. A perfect AI-generated image would be nothing more than a polished stock image – ultimately, it would cease to be art.

Future Research Agenda

The reflections that have been developed in this chapter seem to indicate that a shift in the anthropocentrism of art is occurring: once an essentially human activity, art is now no longer just human but Human^n, in the sense argued. This could lead to the worry that the present account can make art even more anthropocentric, a consequence that not everyone may welcome. However, while enhancing the human element of art, the Human^n option acknowledges the influence of AI technology in some of the current artistic processes and does not rule out the possibility of AI progressively acquiring more autonomy from human creators and developing into something different in the future. One of the core aims of this chapter is to provide a reflection which does not solely focus on the current status of the technology but remains open to its developments: given the impressive pace of evolution of AI technology, the option of an "AI artist", which seems preposterous at present, may progressively be approached with the next iterations of generative AI models.

The chapter reflected on the transformation that the increasingly widespread use of AI for the generation of content can bring to the nature and function of art and to the figure of the artist. The art world, however, includes many more actors who have not been considered here and who are crucial in shaping the dynamics of this transformation. Therefore, for a future research agenda, the following three questions are proposed:

RQ1: How should the role of computer scientists and engineers in the landscape of art made with AI be interpreted?

In line with our Human^n proposition, humans are still playing a central role. Engineers, for instance, are behind the creation of the architecture of algorithmic models and the design of the interface through which users can access them. The audience is a key actor as well:

RQ2: To what extent is the role of the audience set to change with the broader adoption of generative AI models?

Notwithstanding the relevance of human actors, the influence of the algorithm cannot be denied, not only in the creation of content but also in its selection, display, and recommendation to the audience.

RQ3: Are algorithms taking over the role that until now has been that of art curators and of the art establishment?

For how long, and to what extent, art will remain human is an open issue that these questions might help to unravel.

References

Anscomb, C. (2022). Creating art with AI. *Odradek: Studies in Philosophy of Literature, Aesthetics, and New Media Theories*, *8*(1), 13–51.

Becker, H. S. (2008). *Art worlds*. University of California Press.

Boutin, V., Fel, T., Singhal, L., Mukherji, R., Nagaraj, A., Colin, J., & Serre, T. (2023). *Diffusion models as artists: Are we closing the gap between humans and machines?* https://arxiv.org/abs/2301.11722.

Campbell, J. (2001). *Thou art that: Transforming religious metaphor*. New World Library.

Candy, L., & Edmonds, E. (Eds.). (2002). *Explorations in art and technology*. Springer.

Copyright Registration Guidance. (2023). *88 federal register 16190 (March 16, 2023), 37 CFR Part 202*. https://www.federalregister.gov/documents/2023/03/16/2023-05321/copyright-registration-guidance-works-containing-material-generated-by-artificial-intelligence

El-Assady, M., & Moruzzi, C. (2022). Which biases and reasoning pitfalls do explanations trigger? Decomposing communication processes in human – AI interaction. *IEEE Computer Graphics and Applications*, *42*(6), 11–23.

Franco, F. (2022). *The algorithmic dimension: Five artists in conversation*. Springer Series on Cultural Computing. Springer. https://doi.org/10.1007/978-3-319-61167-9

Gombrich, E. H. (1995). *The story of art*. Phaidon.

Gunther, Y. H. (2017). The ineffable in art: On what can't be said. *Literature & Aesthetics*, *27*(2), 17–36.

Hailer, M. (2014). Negative theologie [Negative theology]. In M. Hailer (Ed.), *Religionsphilosophie* [*Philosophy of religion*] (pp. 42–54). Vandenhoeck & Ruprecht.

Hertzmann, A. (2018). Can computers create art? *Arts*, *7*(18). https://doi.org/10.3390/arts7020018

Kaufman, A. B., & Kaufman, J. C. (2015). *Animal creativity and innovation*. Academic Press.

Kuperus, G. (2016). The self as a becoming work of art in early romantic thought. *Idealistic Studies*, *46*(1), 65–77.

Lavoipierre, A. (2022, November 25). A journey inside our unimaginable future. *ABC News*. Background Briefing. www.abc.net.au/news/2022-11-26/loab-age-of-artificial-intelligence-future/101678206

Mazzone, M., & Elgammal, A. (2019). Art, creativity, and the potential of artificial intelligence. *Arts*, *8*(1), 26. https://doi.org/10.3390/arts8010026

Mori, M., MacDorman, K. F., & Kageki, N. (2012). The Uncanny valley [from the field]. *IEEE Robotics & Automation Magazine*, *19*(2), 98–100.

Oppenlaender, J., Linder, R., & Silvennoinen, J. (2023). *Prompting AI art: An investigation into the creative skill of prompt engineering*. https://arxiv.org/abs/2303.13534

Perkins, D. N. (1988). Art as understanding. *The Journal of Aesthetic Education*, 22(1), 111–131.

Radford, A., Kim, J. W., Hallacy, C., Ramesh, A., Goh, G., Agarwal, S., Sastry, G., Askell, A., Mishkin, P., Clark, J., & Krueger, G. (2021). Learning transferable visual models from natural language supervision. In *International conference on machine learning. proceedings of the 38th international conference on machine learning* (pp. 8748–8763). PMLR.

Rombach, R., Blattmann, A., Lorenz, D., Esser, P., & Ommer, B. (2022). High-resolution image synthesis with latent diffusion models. In *Proceedings of the IEEE/CVF conference on computer vision and pattern recognition* (pp. 10684–10695). IEEE.

Royle, N. (2003). *The uncanny*. Manchester University Press.

Sarkar, A. (2023). *Enough with "human-AI collaboration"*. Extended Abstracts of the 2023 CHI Conference on Human Factors in Computing Systems. https://dl.acm.org/doi/abs/10.1145/3544549.3582735

Simon, H. (1985). What we know about the creative process. In R. Kuhn (Ed.), *Frontiers in creative and innovative management* (pp. 3–22). Ballinger.

Tatarkiewicz, W. (2012). *A history of six ideas: An essay in aesthetics*. Springer.

Vallor, S. (2022). The AI mirror: Reclaiming our humanity in an age of machine thinking. In *Proceedings of the 2022 AAAI/ACM conference on AI, ethics, and society*. ACM. https://dl.acm.org/doi/10.1145/3514094.3539567

vanEenoo, C. (2012). The art of making mistakes. *International Journal of Arts*, 2(3), 7–10.

von Kleist, H., & Hamburger, M. (1951). On the gradual construction of thoughts during speech. *German Life and Letters*, *5*, 42–46.

Wan, R., Etori, N., Badillo-Urquiola, K., & Kang, D. (2022). *User or labor: An interaction framework for human-machine relationships in NLP*. https://arxiv.org/abs/2211.01553

Zangwill, N. (1999). Art and audience. *The Journal of Aesthetics and Art Criticism*, *57*(3), 315–332.

Zylinska, J. (2020). *AI art: Machine visions and warped dreams*. Open Humanities Press.

Chapter 8

AI and Creativity in Marketing

Contrasting the Experience and Perspectives of Practitioners and Academics

Pasha Abkenar, Francisco Tigre Moura, and Clare Hindley

Introduction

The generative artificial intelligence (GAI) market is projected to reach US$110.8 billion by 2030, while the latest reports by McKinsey estimate the contribution of GAI to the global market as US$2.6 to US$4.4 trillion annually (Acumen, 2022; Chui et al., 2023).

While many industries strongly embrace GAI tools in the workforce, marketing stands out as a particularly significant adopter (McKinsey and Company, 2022). The industry's use of AI is supported by a Salesforce survey conducted before the launch of ChatGPT 3.5, revealing that 62 per cent of marketing professionals had already incorporated some form of artificial intelligence (AI) into their digital marketing strategies (Salesforce, 2022). The research indicates that AI facilitates marketers in accelerating and enhancing decision-making accuracy, identifying data gaps, analysing data in real time, and customizing user engagement (Stone et al., 2020; Ma & Sun, 2020). However, this represents only a fraction of the extensive data demonstrating AI's utility in marketing (Salesforce, 2023). McKinsey and Company (2022) predict that marketing and sales among customer operations, software engineering, and research and development have the potential to deliver 75 per cent of the total annual value from all GAI use. Industry analysts suggest that GAIs have the potential to influence two core facets of marketing: customer experience and productivity (Deveau et al., 2023). Despite the progress of GAI software adoption in numerous fields, academic research into its beneficial effects, particularly within marketing domains, is limited.

Marketers are still in the early stages of the GAI transformation; however, within the broader scope, the effects of GAI are immense in the creative fields of marketing. For example, Kraft Heinz adopted DALL·E 2 for their "Draw Ketchup" campaign. This campaign revolved around their iconic Heinz 57 ketchup bottle (Heinz A.I. Ketchup – Rethink, 2022). By tasking DALL·E 2 to create ketchup images, the AI's output captured the essence of the Heinz 57 bottle and brand, which Heinz advertised around Canada and the United States on their social media channels. The campaign generated 850 million

DOI: 10.4324/9781003453901-11

impressions worldwide (equivalent to more than 2500 per cent of the media investment), a 38 per cent higher engagement rate compared to previous campaigns, and caused notable media outlets like Fast Company, Bloomberg, TechCrunch, and Forbes to cover the campaign as well (AI Ketchup, 2023). After the success of Kraft Heinz's adoption of GAI, other companies such as Beck's Beer also initiated related campaigns. Becks celebrated their 150th anniversary by releasing a limited-edition beer called "Beck's Autonomous". The entire campaign, from ideation, beer recipe, branding to marketing, was generated by AI prompts (Beck's, 2023). Gartner (2023) predicts that, by 2025, large organizations will synthetically generate 30 per cent of all outbound marketing messages. Moreover, additional research provides evidence that consumers cannot differentiate between human-generated and AI-generated content and that AI-generated or co-created content can rival human-made content (Tigre Moura & Maw, 2021; Tigre Moura et al., 2023), which reinforces the benefits that GAI could provide for early adopters,

This chapter discusses a study exploring and contrasting the experience and perceptions of marketing practitioners and academics on the impact of GAI on creativity. It expands the works of Wingström et al. (2022), and their study contrasting the views of computer scientists and new media artists. Thus, similar to Wingström et al. (2022), the focus of the study is on how participants define creativity, whether the definitions of creativity differ between marketing practitioners and academics, and the role of GAI in the creative process. Finally, the chapter suggests future areas of research to provide further insight into the relationship of AI and creativity in marketing.

Creativity Definitions and Dimensions

"Novel" and "Useful" are two relevant factors often found in most definitions of creativity (Anderson et al., 2014; Mumford, 2003; Parkhurst, 1999) and have become widely recognized as the core of a "standard definition" of creativity (Runco & Jaeger, 2012). Notably, Runco and Jaeger (2012) expand upon this standard definition, emphasizing that creativity demands, additionally, a combination of both *originality* and *effectiveness*. While these two criteria have remained constant in creativity definitions since the 1980s, the understanding of creativity has since then largely expanded and includes different models, such as the four C model by Kaufman and Beghetto (2009), which remains a common method for segmenting creativity into analysable components:

- **Mini-c Creativity:** *The often-overlooked process of a person's original interpretation and ideation during their experiences, actions, or events.*
- **Little-c Creativity:** *Everyday creativity activities that are novel and valuable ideas, solutions, and approaches to personal and practical problems.*

- **Pro-c Creativity:** *The field of creativity that represents developmental and effortful progress past little-c but not at the same level or pass the Big-c creativity.*
- **Big-c Creativity:** *These are exceptional, groundbreaking creativity activities that contribute significantly to various fields.*

However, there is no standard set of dimensions for the interpretation of creativity in individuals, products, or technological innovations. Thus, it is widely acknowledged that creativity is a complex construct influenced by a multitude of factors rather than a singular element (Carson et al., 2005).

This study focuses on Wingström et al.'s (2022), "Five Elements of Creativity" to examine the dimensions of creativity. The model includes five factors: *Actor*, *Space*, *Process*, *Domain*, and *Outcomes*, offering a comprehensive and multifaceted framework for understanding creativity.

- **Actor:** The "Actor" refers to the entity responsible for generating the creative work. Conventional definitions predominantly adopt a human-centric perspective, characterizing the actor as an individual with the capacity for original thought and behaviour. Prior to GAI's advancement, this aligns well with most other seminal works, for example Feldman's (1998), Rhodes's (1961). However, Wingström et al. (2022) acknowledge an emerging shift in academic discourse regarding the definition of an "Actor" in the creative process, signifying a departure from the traditional human-centric paradigm to incorporate a diverse array of actors, now including non-human elements such as artefacts and technological systems.
- **Process:** The creative process should be understood in four stages, as described by Wallas (1926): *preparation*, *incubation*, *illumination*, and *verification*. In the process, Wingström et al. (2022) also acknowledge the roles of *divergent* and *convergent* thinking as core factors enabling creativity. Divergent thinking (mainly related to the *preparation* stage) involves the ability to generate multiple new ideas or solutions to a given problem. Convergent thinking (closely associated with the *illumination* stage) refers to the capacity to synthesize these diverse ideas into a cohesive and effective solution – abilities, which according to the authors, AI could potentially incorporate.
- **Outcome:** According to the definitions of creativity, the result of creative work needs to be original and valuable. However, levels or types of creativity are different for each individual, as defended by Kaufman and Beghetto's (2009) framework. Wingström et al. (2022) acknowledge that the evaluation of an outcome's originality and usefulness remains a challenging subject throughout different domains, with AI's creative capacities often being subjectively judged based on the outcomes they generate and the context in which they are embedded.
- **Domain:** Baer and Kaufman (2005) proposed a hierarchical classification of domains into general thematic areas: *specific* domains and

microdomains. This framework differentiates between domain-general and domain-specific theories of creativity, where the former suggests that creative traits and skills are universally applicable across all domains. At the same time, the latter states that creativity is unique to each domain, requiring domain-specific expertise. This discussion extends to AI, which typically runs domain-specific, being trained with data tailored to particular applications (Anantrasirichai & Bull, 2021).

- **Space:** The dimension of "Space" in the creative process is multifaceted, suggesting that it is an active, interconnected network where the interplay among actors, artefacts, and environments is central to the creative process. This perspective supports the interactionist approach described by Amabile (1983) and aligns with Rhode's (1961) "Press" concept. Wingström et al. (2022) also suggest that incorporating AI into the dimension of "Space" allows us to transcend physical confines, extending into digital domains that foster co-creativity between humans and AI. This perspective highlights the necessity of recognizing "Space" not as a mere backdrop but as a crucial, yet often implicit, element in the definitions of creativity, which according to Wingström et al. (2022), is often overlooked.

Creativity in Marketing

"It is difficult to identify any marketing activities in which creativity could not play an important role" (McIntyre, 1993, p. 36). Ameen et al.'s (2022) extensive review of literature from 1990 to 2021 identifies creativity as a fundamental component in many marketing areas, including advertising, marketing strategy, logo design, product design and development, process and service design, sales, promotions, and the ever-evolving digital marketing sector.

Advertising, in particular, stands out in marketing literature as a field heavily reliant on creativity with authors consistently emphasizing *novelty* as a crucial ingredient for impactful advertising (Sasser & Koslow, 2013; Smith & Xia, 2004). Further studies (e.g. Jin et al., 2019; Shen et al., 2020) find creative advertising more likely to be correctly recognized, recognized faster, and able to weaken competing brands. Reinartz and Saffert's (2013) study of 430 TV advertising campaigns of Germany's 90 fastest-moving consumer goods brands concluded that the most impactful dimension was *elaboration*, followed by Smith and Xia's (2004) added *artistic value* model. *Originality* and *flexibility* were also crucial, along with *synthesis*, another construct suggested by Smith and Xia (2004), which evaluates the integration of seemingly unrelated objects or ideas.

Further studies reinforce the role of creativity in acting as a vital source of competitive advantage for firms and facilitating the adaption, integration, and deployment of resources to align with dynamically changing business environments (Cacciolatti & Lee, 2016; Finch et al., 2013). Moreover,

Turchyn and Ostroverhov (2019) observe a shift in online marketing towards collaborative creativity, which has become a major force driving contemporary digital marketing initiatives.

GAI in Creative Marketing

Vinchon et al. (2023) propose four scenarios for the interplay between GAI and human creativity:

- **Co-cre-AI-tion:** *A symbiotic integration of human and AI efforts, creating a new standardization where collaborative work is essential.*
- **Organic:** *AI replacement of a large portion of careers that can be automated, leaving non-AI-supported creative outputs as more attractive, unique, and authentic.*
- **Plagiarism 3.0:** *Users of GAI tools who intend to appear productive and creative "draw" heavily from such tools without citing the source of inspiration or information.*
- **Shut down:** *Creative work viewed by users as less valuable or visible than an AI creation, referred to as "Creative Mortification".*

Co-cre-AI-tion aligns well with Feldman's (1998) view of collaborative activities in creative tasks, as he challenges the myth of solitary genius by arguing that creativity is the result of cooperation, a supportive environment, and educational contexts. The contribution of GAI to creativity is a subject of ongoing academic debate. As highlighted by Wingström et al. (2022), there is an emergent discussion on redefining the "*Actor*" within the creative process.

Scepticism regarding the creative capability of GAI often centres on Rhodes's (1961) "*Process*" dimension, which emphasizes the need of cognitive activities to reach the final creative outcome. Daikoku (2021, in Tigre Moura et al., 2023), states that AI algorithms are tailored for efficiency and not specifically for creativity suggesting that these systems focus on complying with their encoded program, while creativity seen in humans is sometimes achieved by going against existing rules.

However, as technology advances and bridges the gap in replicating human skills, programming algorithms could mitigate these arguments to be more unexplainable, random, or coincidental, possibly leading to creative outcomes (Jacob, 1995; Loughran & O'Neill, 2020). Likewise, an argument could be made that the standard definitions of creativity and common measurement do not require real-world experience, emotion, or inspiration (Haase & Hanel, 2023). Although the literature covering human against GAI creativity is empirically thin, studies have indicated that when chatbots are asked the same question as their human counterparts, they can generate a larger quantity of ideas that are said to be equal to or more original than the outputs made by humans (Haase & Hanel, 2023). These results go hand in hand with other studies which have indicated that AI outputs are indistinguishable from that of human output (Dwivedi et al., 2023).

So far, the existing literature emphasizes the role of originality in creativity but mainly neglects the equally important aspect of effectiveness. Therefore, GAI's production of original ideas is theoretically likely; however, its meaningfulness and implementation in the creative process remain a human responsibility. Drawing upon the reviewed and available scholarly work, the current consensus leans towards these tools contributing to the creative process as an assistant role for human creativity, regardless of whether they are creatives or non-creatives (Anantrasirichai & Bull, 2021; Beghetto, 2023; Haase & Hanel, 2023; Kowalczyk, 2023; Tigre Moura, 2023).

Despite benefiting from the broad enhancements GAI offers to creative practices, there is a lack of research focused on GAI's direct implications for the marketing industry. Even though authors have called for further research on this subject (Ameen et al., 2022; Dwivedi et al., 2023; Kowalczyk, 2023), these calls have been largely ignored, with few empirical studies conducted to date. Instead, contemporary literature has tended to focus on speculative discussions involving GAI's potential to augment marketing creativity.

Another important point in the discussion of GAI is the limitations and ethical implications. Among text-to-image generative adversarial network software, the primary focus regarding the ethical implications is related to ownership and copyright infringement (Epstein et al., 2023; Brown, 2021). The system's apparent tendency to deliberately stitch various existing contents together leads to the complexity of identifying the rightful "*Actor*" behind the creative output. A significant challenge in formulating regulations to address these issues may come from the human tendency to anthropomorphize GAI tools (Epstein et al., 2023).

Other challenges are, for example, "intention guessing", a concept Bloom (1996) and Snapper et al. (2015) identify as crucial for the audience's viewing experience. Unfortunately, users' experiences generating images that stand out from their usual aesthetics have been more challenging than thought, making it difficult for casual viewers to identify the meaning in a creative work (Epstein et al., 2023). This similarity issue is not confined to visual media only but extends to textual content generated by GAI chatbots. Moreover, Kowalczyk (2023) highlighted concerns within the marketing community regarding the potential for outcomes based on preselected trained data. Another widely recognized problem is the tendency of chatbots to produce "hallucinated" content, which refers to outputs that are coherent sounding but not factually accurate (Brynjolfsson et al., 2023). Such inaccuracies can lower customer trust and harm a company's reputation if not carefully monitored and corrected (Kowalczyk, 2023).

In view of the outlined scenario, the next section discusses the methodology applied to collect the primary data to contrast the experiences and attitudes of marketing academics, and practitioners concerning the impact of GAI on creativity in marketing.

Methodology

To contrast the experience and perceptions of marketing academics and practitioners, the study followed a similar methodology as the work of Wingström et al. (2022). Thirty semi-structured thematic interviews were conducted. The sample included 20 practitioners from a wide spectrum of industries, including finance, logistics, real estate, beauty, and cosmetics, occupying various roles, from solo entrepreneurs and interns to marketing executives. The sample of academics consisted of ten university professors, lecturers, and PhD students providing an academic perspective on the integration of GAI in marketing. This sample size was instrumental in achieving data saturation and the varied backgrounds of participants allowed for a rich tapestry of responses, providing a well-rounded view of the intersection between GAI and creativity in the marketing domain. The data collection was conducted in November 2023 mainly through online interviews lasting 30–50 minutes.

The questions were structured in three main sections: (1) *knowledge and experience with GAI and creativity*; (2) *opinions and experience of GAI and creativity co-cre-AI-tion*; and (3) *future implications of the impact of GAI on creativity*. To ensure the study's validity, the interview questions were designed to follow those of Wingström et al. (2022), with adaptations made to align with the unique goals of this research. This may allow a future publication to contrast findings from both studies.

Results

Definitions of Creativity

In general, the definitions of creativity described by both practitioners and academics (and similar to Wingström et al. [2022]) contained more than one element of creativity. In fact, all five elements of creativity were represented in the responses, either explicitly or implicitly with varying emphasis placed on each. Importantly, "Outcome" and "Process" were mentioned most frequently. A description of each creativity element may be summarized as follows:

Actor: For both groups of respondents, creativity was intrinsically linked to human qualities such as the ability to "*feel*", the utilization of "*self-thoughts and emotions*", and the ability to "*express*". Among these keywords, most definitions referred to the "Actor" with terms such as "*person*", "*myself*", or "*someone*". Notably, no suggestions indicated the "Actor" as any entity that could be deemed as "*non-human*".

Process: "Process" element was frequently characterized by terms that underscore active engagement, such as the combination of knowledge into novel outcomes, finding innovative solutions that address problems more efficiently, and turning existing ideas into something new.

Along with divergent thinking, convergent thinking was also evident in many descriptions. Similar to the "Actor" findings, very few participants suggested GAI's involvement in definitions of creativity. The only suggestions were for the *co-cre-AI-tion* of AI and humans to accomplish a creative outcome.

Outcome: Novelty was mentioned in many iterations. Example of terms included *unique*, *new*, *unusual*, *nonstandard*, *freshness*, *twist*, *innovation*, *wow-effect*, *unexpected*, and *disruptive*. While integral to the utility aspect of creative outcomes, practicality was mentioned less often than novelty, with indication of its presence only suggested through the "*[a]bility to find innovative solutions that address problems more efficiently*". Those who did provide specifics tended to emphasize the personal nature of creativity, preferring a "Little-c" creativity, which suggests personal creative contributions as opposed to universally recognized "Big-c" achievements.

Domain: Overall, participants were split on whether creativity is domain-specific or domain-general. Some participants stated that creativity transcends specific fields and takes a more universal and personal characteristic.

Space: Here particular suggestions indicated that creativity occurs in a network of interaction between actors, artefacts, and environments. For instance, a minor trend within the definitions suggests that creativity in marketing needs to function within certain constraints and regulatory frameworks. Additionally, a suggestion was made that moral compliance also shapes the space for creativity, and the concept of collaboration was also mentioned referring to the use of AI to enhance the creative process.

Differences in Creativity Definitions between Academics and Practitioners

The analysis of definitions shows that academics often approach the subject of creativity from a theoretical standpoint, while practitioners often frame creativity within the context of its practical application.

Inner Expression versus Strategic Creation: More than half of the practitioners directly or indirectly suggested that certain human characteristics are intrinsic to the creative process. Creativity for practitioners seems to be driven by their inner expressions and skills. In contrast, academics did not refer to their inner expressions influencing their creative works. Instead, many emphasized a goal-oriented approach towards a more structured and strategic form of creativity.

Boundary Exploration versus Rule Adherence: The approach of academics to creativity adhered to established boundaries, specifically within the strategic boundaries of company objectives and consistently mentioned structured innovation, balancing the novel with the necessary and the imaginative with the implementable (as potentially supported by the current literature on the topic). Practitioners, contrarily, saw creativity as thriving in a boundary-free and individualistic expression of ideas. This was supported by the connection of creativity with other art forms, that is music, art, and poetry.

Problem-Solving versus Innovative Evolutions: Many practitioners stated that creativity involves a great deal of problem-solving, of "*looking at problems and tasks from different angles [and] challenging the general approach*". This is dynamic, centred on the practical application of ideas to overcome obstacles rather than solely focusing on the evolution of existing concepts. From the academic perspective, more than half of the respondents explained that creativity involves refinement and reinterpretations of existing ideas rather than creating solutions for immediate problems. This process of "*the combination or recombination of existing thoughts and opinions*" was considered crucial in developing something new.

Should Creativity Be Redefined in the Era of GAI?: Most practitioners (12) stated that while the tools and methods within the creative process may evolve due to GAI, the fundamental essence of creativity should not change. Creativity is intrinsically a human attribute and GAI should be considered a tool to enhance human creativity. Others defended an expanded understanding of creativity due to the rising collaborative dynamics between humans and GAI. GAI is seen as a partner in the creative process, capable of generating novel ideas and enhancing the creative output. Among academics, six suggested that redefining creativity might be beneficial in this new era of human and GAI collaboration proposing that *co-cre-AI-tion* includes the ability to generate novel ideas and the skill to utilize GAI tools effectively. Moreover, academics view GAI as an integral part of the creative process, where the combined output of human and GAI collaboration forms a new kind of creativity.

The Role of GAI in the Creative Process

The data collected suggest that GAI has a significant influence among marketing experts. This influence was anticipated from the start, as most participants (28) acknowledged their understanding of GAI without needing further clarification. The remaining participants also recognized the term after being shown examples of GAI applications. All participants, except one, explained that they have already adopted one or more generative tools in their current work process.

GAI Tool Adoption among Practitioners and Academics: The analysis indicates a prominent preference for ChatGPT among academics and practitioners in the marketing field, with an adoption rate of 9:10 for academics and 17:20 for practitioners. However, the utilization of GAI tools in marketing manifests more differently between practitioners and academics than adoption rates. For marketing practitioners, GAI tools play a pivotal role in the early stages of projects, especially in brainstorming and idea generation. They are primarily engaged in their tangible marketing outputs. For example, creating engaging material, fine-tuning language for diverse audiences, and automating repetitive tasks to optimize and streamline their creative process. In contrast, marketing academics use GAI tools more for conceptual and strategic purposes. While the adoption and functional application of GAI tools are outlined, there remains a debate as to whether the output of these tools could constitute creativity.

Can GAI Be creative?: The majority of interviewees believe that while GAI exhibits creative potential, it is "*not creative on its own*" but rather creative when paired with a human counterpart.

Most marketing experts (61 per cent) supported co-creativity by describing collaboration with GAI. Some respondents directly linked GAI with words such as "*partner*", "*co-working partner*", "*50/50 partners*", or "*partnership*". Other respondents disagreed with the concept of an independently creative AI and agreed that GAI helps enhance or assist their creative process.

Impact on the Creative Process: Among the 20 practitioners, suggestions indicate that GAI has significantly impacted the creative process in marketing, particularly in the early and final stages, and in enhancing overall speed and efficiency. It swiftly generates ideas and structures, reducing the time spent on initial phases allowing marketers to focus on more complex aspects of projects. Likewise, among academics, GAI tools are primarily praised for their role in the initial stages of creativity, such as brainstorming and ideation. Similarly, GAI tools could provide ideas and results beyond what one might achieve independently and help cross-reference initial ideas and generate additional information that could complement the creative process. This mirrors a peer review, although in a preliminary and automated form.

Drawbacks and Challenges of GAI: Along with the positive contribution of GAI for marketing experts, there is also a multifaceted range of challenges and drawbacks, reflecting both operational complexities and deeper concerns about the nature of creativity. A primary problem among these challenges is the learning curve involved in effectively interacting with GAI. Many experts mentioned that they have encountered difficulties forming the proper prompts to elicit the desired responses, as a GAI tool's output largely depends on the quality of the input it receives. In terms of practicality, this has occasionally proven more time-consuming for some than manual creation. This inefficiency is further compounded by GAI's tendency to "cycle" between ideas, even after attempts to dismantle the thought process, leading some practitioners to express worries that widespread use of similar GAI tools could lead to uniformity in creative outputs. Moreover, technical limitations and inaccuracies of GAI pose significant challenges for some. For instance, one practitioner raised concerns from her experience of GAI reinforcing stereotypes or biases and occasionally hallucinating facts or connections. Furthermore, nine practitioners expressed concerns about interacting with GAI too much with regard to company data. Additionally, although not frequently discussed, the concept of demotivation among creative professionals was raised, specifically among practitioners encouraged to adopt such tools in the workforce. This included its potential to devalue human creativity, lower quality standards, job insecurity, and general human demotivation. The convenience of GAI might discourage more profound creative efforts, while recognizing GAI's utility in alleviating mundane tasks and acknowledging that it could lessen the enjoyment and personal connection in creative work.

Conclusion

The findings of the study presented in this chapter contribute to the ongoing debate on the impact of GAI on creativity. While Wingström et al.'s (2022) research primarily assessed perspectives of computer scientists and new media artists, this study expands to account for the field of marketing, contrasting academics and practitioners.

The primary data showed definitions of creativity which resonate with the standard definitions found in the current literature, emphasizing novelty and efficiency as critical components. A more detailed analysis, however, revealed that while marketing expert perspectives often align with established theoretical frameworks, there are notable disparities, particularly concerning the role of GAI in creative processes. For example, in the "*Actor*" element, the consensus among participants supports a human-centric approach differing from the growing AI-led view presented by Wingström et al. (2022). Similarly, the "*Process*" and "*Space*" elements, in contrast to the literature suggesting an expanding role for GAI in creative domains, did not much reflect on GAI's involvement. This preference for a human-centric approach does not, however, necessarily suggest resistance to evolving the definition of creativity in the era of GAI advancement. The responses to this issue are diverse and divided. Some experts argue for preserving the traditional definition of creativity, which emphasizes human expression and traits. Others advocate for redefining creativity to reflect the emerging cooperative dynamics between humans and GAI, a notion that is also gaining traction in contemporary literature.

Many participants expressed doubts about GAI's capability for independent creativity, reflecting similar concerns found in the literature regarding the necessity for cognitive processes, intentionality, and emotional depth to be present within creativity. Thus, the findings of this study confirm the literature's view that GAI currently functions best as an augmentative force in the creative process rather than an independent creative entity.

Furthermore, regarding *Co-cre-AI-tion*, additional similarities were found between interview findings and the reviewed literature. Specifically, the fact that co-creating with GAI could minimize time spent in certain areas of the creative process, allowing marketers to focus on other creative tasks. Similarly, the perspective of academics aligned with the literature findings on the increased output quality of writing while lowering the time spent on tasks.

Finally, the interview data contribute to the existing literature by providing tangible examples of how GAI tools are actively shaping the creative marketing process. Operational inefficiencies such as the steep learning curve required for prompt engineering, a risk of producing homogeneous and biased outputs, and a fear of data misuse sharply contrast with the literature's focus on ethical issues, for example copyright infringement. However, concerns over the homogenization of creative content and the production of "hallucinated" content resonate with the literature's discussion of the lack of uniqueness and factual inaccuracies in AI-generated works. Beyond confirming these technical limitations, the interviews provide an additional human dimension to the conversation, highlighting practitioners' worries about job security, the potential devaluation of human creativity, and the risk of demotivation due to over-reliance on GAI tools – issues that are less emphasized in existing scholarly work, even though motivation has been mentioned within the dimensions of creativity.

Future Research Agenda

Findings presented in this chapter, along with the previous discussion involving the current literature on AI and creativity, raise a series of relevant questions. These questions involve, for example, the understanding of the "*Actor*" in the creative process, the influence of GAI on creative diversity, ethical considerations, differences in perspectives between individuals from different creative domains (academic and practitioners), and potential long-term impacts on motivation. These questions aim to deepen our comprehension of the relationship between human creativity and AI, which will hopefully help us further understand how GAI is shaping the evolving landscape of creativity within the marketing domain. Here are three questions to be addressed in future research:

RQ1: How should the concept of the "*Actor*" in the creative process be redefined to account for the evolving role of GAI and its increasing influence in the marketing domain?

The study's findings indicated a human-centric approach in defining the "*Actor*" element in creativity, contrasting with the growing view encompassing AI in the current literature. Thus, future research should focus on examining the evolving roles and contributions of both human and non-human entities within the creative process to develop a more up-to-date and holistic understanding of the "*Actor*". Future research should focus on reconceptualizing the "*Actor*" element to reflect the collaborative dynamics between humans and GAI, considering diverse perspectives and acknowledging the multifaceted nature of creativity in marketing.

RQ2: To what extent does the integration of GAI in the creative marketing process impact the diversity and uniqueness of creative outputs, and how should practitioners effectively navigate potential challenges related to homogeneity?

The chapter also revealed concerns (of practitioners) regarding the risk of producing homogeneous and biased outputs with GAI tools. Thus, future research should investigate strategies to mitigate homogeneity, ensuring that the integration of GAI enhances creativity without compromising uniqueness. After all, as mentioned often in this chapter, *novelty* is a fundamental factor in the definition and understanding of creativity.

RQ3: In what ways should the ethical considerations and limitations related to GAI in creative marketing practices be addressed to ensure responsible and sustainable use?

Finally, the study also revealed potential challenges towards the use of GAI in marketing, such as data misuse, biases, and the learning curve associated with GAI. Thus, future research should further explore frameworks and guidelines for responsible AI use in marketing, considering issues such as copyright infringement, unintentional biases, and the potential impact on job security. Addressing this question will help contribute to establishing novel ethical codes for GAI in the creative marketing domain.

References

Acumen. (2022). *Generative AI market size, share and forecast – 2030*. Acumen Research and Consulting. www.acumenresearchandconsulting.com/generative-ai-market

A.I Ketchup. (2023, May 6). *Ads of the world™* [Video]. www.adsoftheworld.com/campaigns/a-i-ketchup

Amabile, T. M. A. (1983). *The social psychology of creativity: A componential conceptualization*. American Psychological Association. https://psycnet.apa.org/record/1984-06764-001

Ameen, N., Sharma, G. D., Tarba, S., Rao, A., & Chopra, R. (2022). Toward advancing theory on creativity in marketing and artificial intelligence. *Psychology & Marketing*, *39*(9), 1802–1825. https://doi.org/10.1002/mar.21699

Anantrasirichai, N., & Bull, D. (2021). Artificial intelligence in the creative industries: A review. *Artificial Intelligence Review*, *55*(1), 589–656. https://doi.org/10.1007/s10462-021-10039-7

Anderson, N., Potocnik, K., & Zhou, J. (2014). Innovation and creativity in organizations. *Journal of Management*, *40*(5), 1297–1333. https://doi.org/10.1177/0149206314527128

Baer, J., & Kaufman, J. C. (2005). Bridging generality and specificity: The amusement park theoretical (APT) model of creativity. *Roeper Review*, *27*, 158–163.
Beck's. (2023). Pre-order | Beck's autonomous. *Becks.de*. https://becks.co.uk/#signup
Beghetto, R. A. (2023). A new horizon for possibility thinking: A conceptual case study of human × AI collaboration. *Possibility Studies & Society*, *1*(3). https://doi.org/10.1177/27538699231160136
Bloom, P. (1996). Intention, history, and artifact concepts. *Cognition*, *60*(1), 1–29. https://doi.org/10.1016/0010-0277(95)00699-0
Brown, A. (2021). Creative partnerships with technology: How creativity is enhanced through interactions with generative computational systems. *Proceedings of the AAAI Conference on Artificial Intelligence and Interactive Digital Entertainment*, *8*(4), 14–20. https://doi.org/10.1609/aiide.v8i4.12555
Brynjolfsson, E., Li, D., & Raymond, L. (2023). *Generative AI at work*. National Bureau of Economic Research. https://doi.org/10.3386/w31161
Cacciolatti, L., & Lee, S. H. (2016). Revisiting the relationship between marketing capabilities and firm performance: The moderating role of market orientation, marketing strategy and organisational power. *Journal of Business Research*, *69*(12), 5597–5610. https://doi.org/10.1016/j.jbusres.2016.03.067
Carson, S. H. C., Peterson, J. B. P., & Higgins, D. M. H. (2005). Reliability, validity, and factor structure of the creative achievement questionnaire. *Creativity Research Journal*, *17*(1), 37–50.
Chui, M., Hazan, E., Roberts, R., Singla, A., Smaje, K., Sukharevsky, A., Yee, L., & Zameel, R. (2023). *The economic potential of generative AI: The next productivity frontier*. McKinsey & Company. www.mckinsey.com/capabilities/mckinsey-digital/our-insights/the-economic-potential-of-generative-ai-the-next-productivity-frontier#introduction
Daikoku, T. (2021). Discovering the neuroanatomical correlates of music with machine learning. In: Miranda, E.R. (Ed.), *Handbook of artificial intelligence for music: Foundations, advanced approaches, and developments for creativity* (pp. 117–161). Springer, Cham. https://doi.org/10.1007/978-3-030-72116-9_6
Deveau, R., Joseph Griffin, S., & Reis, S. (2023, May 11). *AI-powered marketing and sales reach new heights with generative AI*. McKinsey & Company. www.mckinsey.com/capabilities/growth-marketing-and-sales/our-insights/ai-powered-marketing-and-sales-reach-new-heights-with-generative-ai
Dwivedi, Y. K., Kshetri, N., Hughes, L., Slade, E., Jeyaraj, A., Kar, A. K., Baabdullah, A. M., Koohang, A., Raghavan, V., Ahuja, M., Albanna, H., Albashrawi, M. A., Al-Busaidi, A. S., Balakrishnan, J., Barlette, Y., Basu, S., Bose, I., Brooks, L., Buhalis, D., . . . Wright, R. (2023). Opinion paper: "So what if ChatGPT wrote it?" Multidisciplinary perspectives on opportunities, challenges and implications of generative conversational AI for research, practice and policy. *International Journal of Information Management*, *71*, 102642. https://doi.org/10.1016/j.ijinfomgt.2023.102642
Epstein, Z., Hertzmann, A., Akten, M., Farid, H., Fjeld, J., Frank, M. R., Groh, M., Herman, L., Leach, N., Mahari, R., Pentland, A., Russakovsky, O., Schroeder, H., & Smith, A. (2023). Art and the science of generative AI. *Science*, *380*(6650), 1110–1111. https://doi.org/10.1126/science.adh4451
Feldman, D. (1998). *The development of creativity* (pp. 169–186). Cambridge University Press, eBooks. https://doi.org/10.1017/cbo9780511807916.011
Finch, D., Nadeau, J., & O'Reilly, N. (2013). The future of marketing education: A practitioner's perspective. *Journal of Marketing Education*, *35*(1), 54–67. https://doi.org/10.1177/0273475312465091

Gartner. (2023). *Generative AI: From human, to machine, to artificial intelligence* [Video]. Gartner. https://www.gartner.com/en/topics/generative-ai
Haase, J. H., & Hanel, P. H. (2023, March 21). Artificial muses: Generative artificial intelligence chatbots have risen to human-level creativity. *Journal of Creativity*, *33*(3). https://doi.org/10.1016/j.yjoc.2023.100066
Heinz A.I. Ketchup – Rethink. (2022). *Rethink*. https://rethinkideas.com/work/heinz-ketchup/2022/digital-social/heinz-a-i-ketchup/
Jacob, B. J. (1995, June 11). *Composing with genetic algorithms*. International Computer Music Conference, Banff, AB.
Jin, H. S., Kerr, G., & Suh, J. (2019). Impairment effects of creative ads on brand recall for other ads. *European Journal of Marketing*, *53*(7), 1466–1483. https://doi.org/10.1108/ejm-10-2017-0674
Kaufman, J. C., & Beghetto, R. A. (2009). Beyond big and little: The four c model of creativity. *Review of General Psychology*, *13*(1), 1–12.
Kowalczyk, P. K. (2023). *Nudging creativity in digital marketing with generative artificial intelligence: Opportunities and limitations*. AIS Electronic Library (AISeL). https://aisel.aisnet.org/ecis2023_rip/22/
Loughran, R., & O'Neill, M. (2020). Evolutionary music: Applying evolutionary computation to the art of creating music. *Genet Program Evolvable Mach*, *21*, 55–85. https://doi.org/10.1007/s10710-020-09380-7
Ma, L., & Sun, B. (2020). Machine learning and AI in marketing – connecting computing power to human insights. *International Journal of Research in Marketing*, *37*(3), 481–504. https://doi.org/10.1016/j.ijresmar.2020.04.005
McIntyre, R. P. M. (1993). *An approach to fostering creativity in marketing*. Taylor & Francis. www.tandfonline.com/doi/abs/10.1080/10528008.1993.11488394
McKinsey & Company. (2022, December 6). *The state of AI in 2022 and a half decade in review*. www.mckinsey.com/capabilities/quantumblack/our-insights/the-state-of-ai-in-2022-and-a-half-decade-in-review
Mumford, M. D. (2003). Where have we been, where are we going? Taking stock in creativity research. *Creativity Research Journal*, *15*(2–3), 107–120. https://doi.org/10.1080/10400419.2003.9651403
Parkhurst, H. B. (1999). Confusion, lack of consensus, and the definition of creativity as a construct. *Journal of Creative Behavior*, *33*(1), 1–21. https://doi.org/10.1002/j.2162-6057.1999.tb01035.x
Reinartz, W. J. R., & Saffert, P. S. (2013, June 1). Creativity in advertising: When It works and when it doesn't *Harvard Business Review*, *91*.
Rhodes, M. R. (1961). *An analysis of creativity on JSTOR*. www.jstor.org/stable/20342603
Runco, M. A. R., & Jaeger, G. J. J. (2012, February 10). *The standard definition of creativity*. Taylor & Francis. www.tandfonline.com/doi/abs/10.1080/10400419.2012.650092
Salesforce. (2022, August 8). *State of marketing report* (8th ed.). www.salesforce.com/resources/research-reports/state-of-marketing/
Salesforce. (2023). New research: 60% of marketers say generative AI will transform their role, but worry about accuracy. *Salesforce*. www.salesforce.com/news/stories/generative-ai-for-marketing-research/
Sasser, S. T. S., & Koslow, S. K. (2013). *Desperately seeking advertising creativity: Engaging an imaginative*. Taylor & Francis. www.tandfonline.com/doi/abs/10.2753/JOA0091-3367370401
Shen, W., Liu, Z., Ball, L. J., Huang, T., Yuan, Y., Bai, H., & Hua, M. (2020). Easy to remember, easy to forget? The memorability of creative advertisements.

Creativity Research Journal, *32*(3), 313–322. https://doi.org/10.1080/10400419.2020.1821568

Smith, R. E., & Xia, Y. (2004). Toward a general theory of creativity in advertising: Examining the role of divergence. *Marketing Theory*, *4*(1–2), 31–58. https://doi.org/10.1177/1470593104044086

Snapper, L., Oranç, C., Hawley-Dolan, A., Nissel, J., & Winner, E. (2015). Your kid could not have done that: Even untutored observers can discern intentionality and structure in abstract expressionist art. *Cognition*, *137*, 154–165. https://doi.org/10.1016/j.cognition.2014.12.009

Stone, M., Aravopoulou, E., Ekinci, Y., Evans, G., Hobbs, M., Labib, A., Laughlin, P., Machtynger, J., & Machtynger, L. (2020). Artificial intelligence (AI) in strategic marketing decision-making: A research agenda. *The Bottom Line: Managing Library Finances*, *33*(2), 183–200. https://doi.org/10.1108/bl-03-2020-0022

Tigre Moura, F. (2023). Artificial intelligence, creativity, and intentionality: The need for a paradigm shift. *Journal of Creative Behavior*, *57*, 336–338. https://doi.org/10.1002/jocb.585

Tigre Moura, F., Castrucci, C., & Hindley, C. (2023). Artificial intelligence creates art? An experimental investigation of value and creativity perceptions. *Journal of Creative Behavior*, *57*(4), 534–549. https://doi.org/10.1002/jocb.600

Tigre Moura, F., & Maw, C. (2021). Artificial intelligence became Beethoven: How do listeners and music professionals perceive artificially composed music? *Journal of Consumer Marketing*, *38*(2), 137–146.

Turchyn, L. T., & Ostroverhov, V. O. (2019, January 1). Modern trends of internet marketing. *Regional Aspects of Productive Forces Development of Ukraine*, *1*(24), 75–85. https://doi.org/10.35774/rarrpsu2019.24.075

Vinchon, F., Lubart, T., Bartolotta, S., Gironnay, V., Botella, M., Bourgeois-Bougrine, S., Burkhardt, J., Bonnardel, N., Corazza, G., Glăveanu, V. P., Hanson, M. H., Ivcevic, Z., Karwowski, M., Kaufman, J. C., Okada, T., Reiter-Palmon, R., & Gaggioli, A. (2023). Artificial intelligence & creativity: A manifesto for collaboration. *Journal of Creative Behavior*, *57*(4), 472–484. https://doi.org/10.1002/jocb.597

Wallas, G. W. (1926). *The art of thought* [Archiev.org]. Harcourt, Brace and Company. https://archive.org/details/theartofthought/mode/2up

Wingström, R., Hautala, J., & Lundman, R. (2022). Redefining creativity in the era of AI? Perspectives of computer scientists and new media artists. *Creativity Research Journal*, 1–17. https://doi.org/10.1080/10400419.2022.2107850

Chapter 9

On the Test Bench

How to Evaluate the Quality of AI Systems?

André Hollstein

Introduction

Quality plays an important role in business, building trust and customer satisfaction while driving profitability, growth, and development (Cvjetković et al., 2021). Whereas in the twentieth century and before quality of work, and thus the outcome of a product or service, was heavily dependent on the performance of employees in companies, their skills, competencies, and ability to cooperate, this situation changed with the widespread usability of artificial intelligence (AI). In the twenty-first century, collaboration between humans and AI plays an increasingly important role. In particular, the capabilities of AI support appear to be able to improve the quality of work outcomes in a variety of ways. A 2023 study by McKinsey shows that the explosive growth of generative AI tools means that nearly one-quarter of surveyed C-level executives say that they are personally using AI tools on a regularly base (Chui et al., 2023). Another study conducted by Boston Consulting Group in 2023 among 13,000 respondents (from management to employees) in 18 countries shows that the use of AI has increased from 22 per cent in 2018 to 50 per cent in 2023 (Beauchene et al., 2023).

But it is not only the range of potential use cases that has expanded, AI also means that the scope and creativity has expanded and does not lead to the end of human employment, as a study in Germany has elaborated (Fregin et al., 2020).

Quality improvement refers to the characteristics or properties of the results of collaboration between human and AI and the value that can be derived from the benefits or performance of this collaborative work process. In other words, the higher the quality of the result, the higher the benefit for the individual or the company. In this sense, the expectation of the result can refer to aspects such as accuracy, efficiency, reliability, or satisfaction with the result of the cooperative work. As will be explained later in this chapter, this is also in line with common definitions of quality (e.g. Besterfield, 2011).

However, quality improvements are not limited to the process of co-creating documents, images, or videos. The efficiency gains through automation (e.g.

DOI: 10.4324/9781003453901-12

reduction of human errors) or the improvement of security aspects in software development, to name just two examples, demonstrate the benefits of collaborative work processes.

Thus, it is important to understand that the quality outcome depends on both AI solution and humans. However, relying on AI systems can lead to misjudgements, which underline the need for human oversight (Dumitru & Halpern, 2023). Furthermore, human skills such as creativity or critical thinking, with its cognitive and emotional dimension, remain important remain important (Al Marri & Obaid, 2023). However, AI solutions can be used as a supporting tool to improve the outcome.

This chapter discusses the importance of quality in daily work processes, like decision-making, problem-solving, process support, or other aspects; what quality means in relation to AI; and how defined quality dimensions can be used to evaluate AI solutions. A quality matrix based on the work of Garvin (1987) is presented to assess the quality of AI in the holistic way. The chapter concludes with a case from software development to illustrate how the quality of an AI-generated output can be assessed.

What is Quality in General?

In the context of an increasingly digital and disruptive world, customer requirements for products and services have changed significantly, as the price/performance ratio, quality and other criteria have become more and more transparent. This transparency provided primarily by the internet and social media allows consumers to take a more active role in influencing product pricing and quality or imposing sanctions based on expectations not being met (Alsubagh, 2015; Rezabakhsh et al., 2006).

For companies, this means meeting the ever-increasing demand to deliver products and services that meet the needs of the market while providing a high level of quality. All of this puts competitive pressure on traditional business models and create a massive need for change in traditional companies due to rapidly changing markets (Verhoef et al., 2021). However, quality is not a new concept in the business world. As early as the nineteenth century, companies were concerned about the quality of their products and tried to improve it by taking appropriate measures (Evans & Lindsay, 2013). Quality assessment and improvement is important because it results in products and services that meet customer perceptions and build trust, while from an internal perspective it increases efficiency and reduces waste and external failures (Luthra et al., 2020; Owen, 2020).

To support and improve quality, numerous standards and quality frameworks provide support. The quality standards and quality frameworks are, for example, the standards of the ISO 9000 series (International Organization for Standardization, 2023) or the Total Quality Management Framework (Luthra et al., 2020).

There are numerous definitions of quality in the literature, or as Foster and Gardner stated, "[i]f you ask 10 people to define quality, you will probably get 10 different definitions" (Foster & Gardner, 2022, p. 4). What many authors agree on is that product or service quality is primarily about meeting the expectations of customers or users (Kumar, 2016). For this reason, there are a variety of definitions that take customer expectations into account. Besterfield (2011, p. 6) summarizes quality in a very generic form of a simple equation: Accordingly, Quality (Q) can be considered as the ratio of Performance (P) to Expectation (E): $Q = P/E$.

However, it is not only important to understand the relationship between quality and consumer expectation, but also to describe how quality can be assessed, as shown by the following definition (Smith, 1993, p. 237): "*Quality is the goodness or excellence of something. It is assessed against accepted standards of merit for such things and against the interests/needs of users and other stakeholders*".

Accordingly, it is not only a question of what quality is but also how it is measured and evaluated. This raises the question of how and by what criteria or standards quality can be assessed. The answer to this question depends on the "object" of the assessment, for example whether it is a product or a service, and on the perspective adopted, for example the user's or the producer's perspective (Dahler-Larsen, 2019). The following sections will explore what quality means in the context of AI solutions and how it can be measured for AI solutions.

What Does Quality Imply in Relation to AI?

The use of AI in business is not a new approach and existed before modern solutions like Stable Diffusion (stability.ai) or ChatGPT (openai.com) made it available to everyone (Delipetrev et al., 2020). However, the special momentum did not come from technical progress, but from the combination of powerful technology, dialogue-oriented usability, and useful results. In addition, these solutions offer interfaces, in the form of lightweight application programming interfaces, that enable many companies to integrate these products into their daily work processes. Furthermore, a whole new ecosystem has formed in which new start-ups are emerging almost daily, using existing or their own AI models to bring solutions to market. In addition, platforms such as Hugging Face (huggingface.co) enable and facilitate experimentation with the possibilities of AI, as they offer a wide range of accessible "models" and "datasets" to test own approaches and develop topic-specific solutions, coupled with moderate prices and little programming effort.

This easy accessibility inspires the creativity of individuals and companies, and at the same time, fuels the concern of falling behind and letting the competition pass. This raises the question of how AI solutions can be used productively to accelerate daily routines, in the sense of "*repetitive, recognizable patterns of interdependent organizational actions carried out by multiple actors*" (Feldman & Pentland, 2003, p. 95), or to support decision-making under complex conditions, in the sense of "*may operate in patterned ways but whose interactions are continually changing*" (Sargut & McGrath, 2011, p. 68).

However, the question is not only whether it is possible to accelerate or support work processes, but also whether AI solutions can deliver better and even higher quality results than humans alone? This raises the question of how to assess the quality of an AI solution and the quality of collaboration? This important aspect will be discussed in the following section.

What is a Good Quality Benchmark for an AI Solution?

The quality of AI models can be viewed and evaluated from various perspectives. In recent years, a large number of benchmarks have been introduced to evaluate the development of AI models. Ott and his colleagues for example have found 3,765 benchmarks for image and natural language processing (Ott et al., 2022). However, less than one-third of these benchmarks were used in different studies.

Benchmarks essentially try to measure the quality of the result, respectively the performance, normally related to specific tasks. Therefore, benchmarks typically contain one or more datasets and metrics for measuring performance (Naveed et al., 2023). However, the results of the benchmarks are highly dependent on the quality of the datasets created for training and evaluation (Holzinger et al., 2023; Naveed et al., 2023).

Nevertheless, evaluating or benchmarking the performance of an AI solution based on its output performance may not be sufficient. This may be due to a lack of comparability, as the complexity and variety of tasks in areas such as computer vision, predictive analytics, and robotics can vary significantly. Furthermore, the occurrence and interpretability of the results are not always given, an aspect that eXplainable AI (XAI) tries to solve (Brandt et al., 2023). Another reason could be that manipulation by the AI system itself happens to create the illusion of good results instead of the desired result (Roger et al., 2023). After all, there is still a big gap between AI pattern recognition and human learning when we need solutions, support humans, or possibly free them from specific tasks (Holzinger et al., 2023; Phung et al., 2023).

In addition to the factors just mentioned, testing an AI solution for performance alone may not be sufficient, as such solutions must always be considered in a broader context when used in a business environment for an intended task. Since AI solutions are essentially software solutions, it makes sense to check in this area whether there are corresponding examples of test procedures that go beyond the performance result.

Quality assessment models for software evaluation exist, for example, in form of the standard ISO/IEC 25010:2011 (ISO/IEC, 2011) and refers to static and dynamic properties of the software itself, which is divided into eight characteristics (Functional Suitability, Performance Efficiency, Compatibility, Usability, Reliability, Security, Maintainability, and Portability) and further 31 sub-characteristics.

If a company does not want to develop AI solutions itself because it does not have the expertise or resources, then it makes sense to look for feasible solutions on the market. In this sense, AI solutions can also be viewed from the perspective of a product or service.

One of the most respected and widely used collections of quality dimensions for analysing the quality of products or services dates to the 1980s. It was published by David Garvin (1987) and describes eight quality dimensions for a strategic analysis of the quality of products or services.

The aspects used by Garvin (1987) can be described as follows:

1. *Performance*: Describes the primary basic properties.
2. *Features*: Defines properties that complement its basic functionality.
3. *Reliability*: Describes the probability that a product or service will malfunction or fail within a specified period.
4. *Conformance*: Defines the degree to which the design and operational characteristics conform to the specified standards or expectation.
5. *Durability*: Criterion for the service life or durability.
6. *Serviceability*: Ease of maintenance, for example simplicity for repairs or improvements.
7. *Aesthetics*: Personal assessment in terms of how a product looks, feels, or is usable.
8. *Perceived Quality*: Describes the personal impression.

As already mentioned, Gavin's work is widely used because the concept of eight dimensions is easily adapted to many cases. Not least because quality is viewed in holistic, value-based, and user-oriented way, it is also used in software development (Riggins & Gregory, 2019). Garvin's concept can therefore also be applied to AI solutions. Against this background, the eight dimensions are explained below using examples:

(1) and (2) **Performance/Features** (separation is specific to the solution and its focus, so these are considered together)

- **Relevance of results:** AI-powered collaboration should lead to solutions that meet the needs and desires of stakeholders.

  Example: If an AI engine is used to translate texts, then the results should be correct in terms of content (including technical terms) and grammar in the chosen language (Wang et al., 2018). *In addition, the engine used should also understand semantics and the varieties of human thought (language jokes, logic, analogies).*

- **Accuracy and precision:** The accuracy and precision of the results are key quality criteria for the use of an AI model. To complete a specific task, a requirement should be fulfilled 100 per cent, even 90 per cent may not be enough, for example, when translating texts or answering customer inquiries through large language models (Gieselmann & Trinkwalder Andrea, 2023; Peinl & Wirth, 2023). There are other examples where inaccuracies in the processing of an AI model can lead to incorrect results with fatal consequences.

  Example: AI solutions for the evaluation of medical images can lead to reduced errors and times in diagnosis and improved performance in the prediction and detection of various diseases (Ghaffar Nia et al., 2023).
- **Reproducibility:** AI models should reliably deliver reproducible results, for instance the results should always be the same or the AI solution comes to the same conclusions.

  Example: An AI solution that evaluates incoming customer emails and decides whether to respond to the concerns with a standardized, automated email or to have them handled by a human should always be available and come to the same, reproducible results.
- **Time efficiency:** Depending on the use of AI-supported processing, it is important that a decision or feedback is made in real time or with low latency adapted from Ghosh (2023). In this context, latency is the amount of time required by an AI model to respond to an input impulse or request, for instance to process a query and return a result.

  Example A: An AI-based recommendation system for an online store should provide real-time product suggestions to prevent the user from leaving the site.

  Example B: An autonomous vehicle needs to be able to react to its environment in real time. When an obstacle is detected on the road, the time it takes for the AI system to process this information and decide is a critical factor.
- **Scalability/Elasticity:** Scalability refers to the ability to effectively increase capacity and performance as requirements grow. Elasticity describes the ability to do this on demand (Arankalle et al., 2020, p. 55). In terms of AI systems, this can mean that a system is able to respond dynamically to the number of user requests. The form of scalability depends on the use case and can refer, for example, to scalability of the data or the model.

  Example: In applications that require real-time responses (e.g. autonomous vehicles or chatbot), the AI system should be able to keep the response time constant despite increasing complexity or load.
- **Continuous learning:** Even before the development and deployment of conversational AI solutions, the desire arose for AI models to be able to take feedback into account and adapt their solutions accordingly. While in the early days of AI research it was mainly the initial learning process (e.g.

supervised, unsupervised, or reinforcement learning) that fed AI models their knowledge, today efforts are made to strive for a continuous learning process, through multi-task learning or transfer learning (Liu, 2017).

Example: In the context of robot control, a robot that has learned to perform a particular task in one environment can use its learned knowledge to perform similar tasks in modified environments or even completely different tasks.

(3) Reliability

- **Robustness:** The robustness of AI models can be viewed from a similar perspective as the reproducibility property. AI models must come to the right decisions under the influence of changing parameters to support humans in a meaningful way.

  Example: An autonomous vehicle should also function reliably under different weather conditions and in unexpected traffic situations. Therefore, it is necessary to drive millions of miles to prove this point (Stilgoe, 2021).
- **Explainable** (also known as "eXplainable AI" or "XAI"): Stakeholders should be able to understand how the AI reaches its conclusions and be influenced by the results. The decisions of complex AI models are often perceived as a "black box" because the results are often perceived as correct, but it is not explainable how the result is reached. Therefore, a separate field of research has emerged to deal with XAI (Saranya & Subhashini, 2023).

  Example: XAI is an important aspect especially when AI systems are used in critical areas such as medicine, finance, and law. Therefore, it is essential to understand how and why certain decisions are made.

(4) Conformity

- **Privacy and ethics:** Privacy and ethical considerations should always be taken into account when using AI, as AI has the potential to deeply interfere in the lives of individuals or groups.

  Example A: The use of facial recognition solutions should not be done without consent or deletion rules, that is AI systems should ask for users' consent or data taken in public spaces, for example, should only be stored for a short time and for a specific purpose. This ensures that user privacy is respected.

  Example B: Anti-bias training: When using or training AI models, it can happen that biases are unconsciously learned from the data. One of the most well-known examples comes from Amazon, whose AI recruitment tool disadvantaged women in hiring because the AI model was trained with a disproportionate amount of data from male employees (Dastin, 2018). *In terms of an AI system, therefore, ensure that all data are considered regardless of gender, ethnicity, or other discriminatory characteristics and that results remain explainable.*

- **Security**: Ensuring the security of AI systems is a multidimensional aspect that covers many other dimensions and refers to measures that ensure that AI solutions are trustworthy, reliable, and free of harmful behaviour or undesirable side effects. This includes aspects such as robustness and reliability as well as traceability and data protection. In addition, there are aspects such as protection against misuse and manipulation. AI models should be protected against attempts to change their functions or behaviour by manipulating the training data or other techniques.

  Example: Especially in the context of autonomous intelligent systems, such as self-driving vehicles or autonomous disease diagnosis, the manipulation of image-based data, by using adversarial examples, can mislead the trained model with respect to the target problem (Aynaci & Köse, 2021). *For example, if an AI model is trained to recognize traffic signs, counterexamples or marginally altered images can cause the AI system to misinterpret one traffic sign as another – for example, a stop sign as a speed limit sign.*

- **Copyrights:** The proper handling of copyrights is also a crucial quality criterion for AI solutions, because non-compliance can lead to legal consequences, including fines, cease-and-desist orders, and possible claims for damages. In addition, an AI solution that respects copyrights increases the trust of its users.

  Example: Very often, it is not clear where the data come from that are used to train AI models. Several copyright lawsuits are pending in the United States. that will have a significant impact on the future of generative AI systems. This is because if the plaintiffs prevail, AI systems in the future will only be able to be trained on data that belong to the "public domain" (Samuelson, 2023). *A case in point is the lawsuit filed against Stable Diffusion for possible infringement of artists' rights* (Butterick, 2023).

(5) Durability

The durability dimension in the context of AI refers to how sustainable, future-proof, and maintainable an AI solution is over time. In the fast-moving field of technology and AI development in particular, the relevance and effectiveness of a solution can quickly diminish, if it is not regularly updated and adapted. Therefore, aspects such as maintainability, learnability, or updateability of the model can form an important quality criterion.

(6) Serviceability

AI solutions are very often AI as a Service (AIaaS) approaches. This shifts the maintenance of such solutions to a third-party provider (Cobbe & Singh,

2021). Therefore, response times, resolution rates of errors, and other aspects that are regulated in contracts or service levels are very important.

(7) Aesthetics

- **Ease of use:** An AI-powered system should have an intuitive, easy-to-use interface and provide easy-to-understand answers.
- **Diversity and inclusion:** A high-quality co-creation process should ensure that a wide range of perspectives and voices are considered. AI models should be trained without bias to ensure that they treat all users fairly.

(8) Perceived Quality

Perceived Quality refers to the subjective assessment and opinion of a user or a group of users regarding to the quality of a product, service, or solution. This may differ from the objective, measurable quality characteristics, must be illuminated by a correct example, and will therefore not be considered further.

It is important to emphasize that the relevance and significance of these criteria can vary depending on the application area and objective. This is especially true against the background of the development of AI systems, on the one hand and the use, from the end-user perspective, on the other. And in many cases, it may also be necessary to find a compromise between different criteria.

Applying Quality Dimensions: The Case of Software Development with Copilot

Why is software development an interesting example? In an increasingly digital world, software plays a central role in all aspects of life. Having high-quality software solutions to manage processes, automate tasks, or communicate with tens of thousands of customers simultaneously is critical to the success of today's businesses. The demand for software is constantly increasing. However, it is not enough to use any software or to develop software without certain rules. In addition to business requirements, software must also meet quality criteria such as security, reliability, and efficiency to ensure the success of the business. Achieving these criteria is not easy and has cost roughly $2.41 trillion in 2022 in the United States alone (Krasner, 2022).

Developing software solutions is first and foremost a multi-layered task that is not always executed error-free by humans. That's because today's software projects are often very complex and can contain millions of lines of code (Desjardins, 2017). The larger and more complex the code becomes, the more difficult it becomes to assure quality. This cognitive load can cause developers to overlook details or errors. Due to the time pressures often present in projects, testing is less thorough or technical debt is accumulated, leading to errors. In addition, not all software projects have adequate test coverage and even more, 100 per cent coverage is not a sufficient condition for good quality (Prause et al., 2017) or an increasing coverage may not translate to higher bug-finding rates (Kochhar et al., 2017).

Many methods have emerged in software development that help to minimize these sources of error, for example code reviews, pair programming, or continuous integration and delivery. Nevertheless, software development remains a challenging task in which developers often operate at the limits of their abilities.

The idea of using AI to support software development is not quite as old as the general idea of AI itself, but dates to the early 1970s (Bobrow & Raphael, 1974). With the development of new generative AI models, new approaches and ideas have emerged on how AI can be used in the process of creating, revising, and improving software to produce simpler, higher quality, and safer code. An example of such is GitHub Copilot (github.com/features/copilot) and is presented below as an example of a co-creation tool in the context of software development.

GitHub Copilot

GitHub Copilot was launched by GitHub, a Microsoft-owned company, in partnership with OpenAI in mid-2021. GitHub sees the development of GitHub Copilot as a new form of pair programming, where instead of the usual two human programmers working on a project to identify bugs and speed up the development process, one is replaced by a virtual partner (Gershgorn, 2021). In general, the literature documents the positive effect of pair programming on quality and a moderately positive effect on duration (Hannay et al., 2009). It can be assumed that this effect is also found in the interaction between a virtual and a human partner, but this has not yet been extensively investigated.

GitHub Copilot is based on Codex, a model that can handle more than a dozen programming languages and interpret simple commands in natural language (Zaremba et al., 2021).

As an AI-powered co-creation tool, GitHub Copilot supports software development quality in several ways:

1. *Creation and completion*: Copilot provides real-time suggestions to create blocks of code, such as for needed functions, or to complete lines of code. This can reduce human errors caused by typos or by overlooking inadequate sections of code. At the same time, requirements can be implemented efficiently, security standards can be met, anti-patterns can be circumvented, and appropriate corrections or improvements can be suggested. A McKinsey study shows software developers can complete coding tasks up to twice faster with generative AI (Deniz et al., 2023).

2. *Documentation and comments*: Copilot can help add comments or provide reminders to complete the documentation, which contributes to the understandability and maintainability of the code.
3. *Best practices*: Copilot can provide recommendations based on best practices and popular coding standards. This can help ensure that the code written is more consistent and compliant to software standards.
4. *Increase productivity*: By quickly completing and suggesting solutions to common programming tasks, Copilot can reduce development time and encourage developers to focus on more complex aspects of their work.

Referring to Garvin's (1987) eight dimensions of quality, the characteristics of GitHub Copilot can be evaluated as follows.

(1) and (2) Performance/Features

This dimension describes the primary, expected basic characteristics of the product. For a tool that supports software development, the following features may be required:

- *Target language*: Copilot should be able to handle the desired programming language, the frameworks that are used, and the libraries that are required.
- *Correctness*: The proposed code should be correct, functional, not repetitive, safe, consistent, and other aspects with the desired result of the developer (Cocca, 2023; Martin, 2008). In addition, Copilot should generate test code to test the developed function directly and independently from the overall code (Beck, 2022; Uddin & Anand, 2019).
- *Speed*: Copilot should respond quickly to input without significant latency and not negatively impact the performance of the development environment or associated processes or systems (e.g. communication with backend systems like databases). In addition, the collaboration and communication should be natural and intuitive.
- *Integration*: GitHub Copilot should integrate with your preferred integrated development environment (IDE) and support built-in features like debugging and deployment workflows, because research shows that IDEs are more than just development tools (Zayour & Hamdar, 2016).

(3) Reliability

The randomness dimension refers to permanently correct function and provision of the application and represents a central criterion for the practical usability and acceptance of the product. In addition to the question of the consistency of the code created or alternative solutions to the same problem, the question of runnability and response time, even in the case of more complex, unclear, or ambiguous requests.

In addition, it is important to consider whether the service is always available or whether the feedback speed varies greatly for the same tasks or does not return runnable examples at all. A feedback mechanism can help to improve the reliability of the tool.

(4) Conformity

Against the backdrop of the conformity dimension, sometimes referred to as "compliance", it is about adherence to standards, guidelines, or specifications. In the context of AI-powered solutions such as GitHub Copilot, this means, among other things, adherence to programming standards, licence compliance (the proposed code is licensed or not from publicly available sources), or consideration of security standards (not using or adapting unsafe programming patterns or practices).

In practice, compliance is difficult to maintain and evaluate by programmers because the source of the code used is usually not transparent and compliance is also difficult to assess. This is because such solutions are used precisely because the application seeks solutions to difficult problems.

(5) Durability

The durability dimension refers to how sustainable, future-proof, and maintainable an AI solution is over the period of use. In the fast-moving field of technology and AI development, in particular, the relevance and effectiveness of a solution can quickly diminish if it is not regularly updated and adapted. Therefore, aspects such as maintainability, learnability, or updateability of the model can form an important quality criterion. In the context of Copilot, as a "Software as a Service" solution, this criterion is not incumbent on the user or the developer, but on GitHub, which provides this cloud solution. Nevertheless, it can represent an important criterion for introducing such a solution into a company as part of the purchasing process.

(6) Serviceability

"Artificial Intelligence as a Service" (AIaaS) aims to make AI solutions available to everyone (Lins et al., 2021). For organizations, cloud-based systems offer the advantage of high availability with low administrative overhead. Conversely, the unavailability of the Copilot is problematic but would not stop overall development, as it could continue without AI support. For AI solutions, serviceability can generally be a buying criterion, as it aims for accessibility, scalability, and on-demand services (Singh et al., 2016).

(7) Aesthetics

GitHub Copilot is designed as an extension for development environments and integrates seamlessly into the code editor and thus does not have its own user interface. The user experience, therefore, consists primarily of the integration into the code editor (seamlessness, intuitiveness, interactivity) but also through the integration of the service into the code editor (simplicity, customizability). Ultimately, the evaluation of this dimension is subjective and can vary from user to user.

(8) Perceived Quality

The perceived quality dimension is analogous to aesthetics, a subjective measure of the extent to which the AI tool, like Copilot, meets or exceeds the expectations of a user or user group. Perceived quality could refer, for example, to the accuracy of suggestions, relevance of suggestions, performance, but communication and support, value for money.

The Role of the User in the Co-Creation Process

Solutions, such as GitHub Copilot, can be a great help in the production process by the fact that it helps to improve the quality of the created software solution. However, it is not guaranteed that the service is adequate, and the result is free of errors. It is also the responsibility of the developers to review the proposals and make sure that they achieve the desired goals.

As the name "Copilot" suggests, it is and remains a supportive tool as many other AI tools. The "pilot" sits in front of the computer. Therefore, it is important that developers don't trust all suggestions of GitHub Copilot, but also review the suggested code critically. Humans bring problem-solving skills, intuition, and contextual understanding. A developer understands the "why" behind a particular implementation, while the AI model is based on pattern recognition and doesn't really "get" what the problem is. The developer sees the big picture and can consider requirements, stakeholder expectations, and end-user needs, while an AI model only knows the context given to it. However, solutions such as GitHub Copilot can help to improve the quality and security of software solutions.

Future Research Agenda

The quality of AI models and solutions is an important criterion for the acceptance and usability of such approaches. Improving quality is an important area of research. How could a more tailored quality assessment model, reflecting the different dimensions of quality, help guide the development of AI models and solutions that are not only powerful but also reliable, ethical, and useful to end users? In addition, there is a lack of concepts that help to fully exploit the potential of AI technologies in interaction with people, especially if AI technology is developing rapidly.

With reference to the example used in this chapter, pair programming with GitHub Copilot, there are also fundamentally interesting questions about quality that should be evaluated further and that can be applied to other examples.

In this sense the following further research questions can be important.

RQ1: What could a quality assessment model tailored more precisely to AI solutions look like?

In this chapter, Garvin's quality model (1987) was applied to evaluate the quality of AI solutions and services. Overall, the evaluation model is very general. This raises the question of how the model can be extended or improved to better reflect the specifics of AI solutions, such as those used in many AI benchmarks (Naveed et al., 2023; Ott et al., 2022).

RQ2: What support concept can help people deal with constant technological change and at the same time realize the potential of AI technologies?

The use of appropriate AI solutions and tools does not unleash the full potential of what is possible, as shown by Deniz and colleagues (2023). Therefore, much emphasis should be placed on training and coaching for AI solutions. However, as the market for solutions is changing rapidly, new challenges are emerging for users. This raises the question of how to provide support that helps people cope with the constant technological change in business, while at the same time unlocking the potential of AI technologies for the business.

RQ3: Could the quality of the results be better through virtual pair programming in terms of security aspects, bug minimization, or other aspects?

Studies have shown that the speed of software development has been greatly improved by the support of AI solutions (Deniz et al., 2023). As Hannay et al.'s study (2009) shows, pair programming generally has a significant effect on outcome quality. However, the question remains whether the quality of the results might also be improved, for example in terms of security aspects or bug minimization.

References

Al Marri, S., & Obaid, H. (2023). The impact of the critical thinking on cognitive building. *Islamic Sciences Journal*, *12*(8), 391–410. https://doi.org/10.25130/jis.21.12.8.17

Alsubagh, H. (2015). The impact of social networks on consumers' behaviors. *International Journal of Business and Social Science*, *6*(1).

Arankalle, C., Dwyer, G., Geerdink, B., Gera, K., Liao, K., & Anand, N. S. (2020). *The artificial intelligence infrastructure workshop: Build your own highly scalable and robust data storage systems that can support a variety of cutting-edge AI applications*. Packt Publishing.

Aynaci, E., & Köse, U. (2021). Manipulation of artificial intelligence in image based data: Adversarial examples techniques. *Journal of Multidisciplinary Developments*, 6, 8–17.

Beauchene, V., de Bellefonds, N., Duranton, S., & Mills, S. (2023, June 7). *AI at work: What people are saying*. Boston Consulting Group. www.bcg.com/publications/2023/what-people-are-saying-about-ai-at-work#

Beck, K. (2022). *Test driven development: By example*. Pearson Education. https://books.google.de/books?id=zNnPEAAAQBAJ

Besterfield, D. H. (2011). *Total quality management* (revised ed.). Pearson.

Bobrow, D. G., & Raphael, B. (1974). New programming languages for artificial intelligence research. *ACM Computing Surveys (CSUR)*, *6*(3), 153–174.

Brandt, R., Raatjens, D., & Gaydadjiev, G. (2023). *Precise benchmarking of explainable AI attribution methods*. http://arxiv.org/abs/2308.03161

Butterick, M. (2023, January 23). *We've filed a lawsuit challenging Stable Diffusion, a 21st-century collage tool that violates the rights of artists*. https://stablediffusionlitigation.com

Chui, M., Yee, L., Hall, B., Singla, A., & Sukharevsky, A. (2023, August 1). *The state of AI in 2023: Generative AI's breakout year*. McKinsey & Company. www.mckinsey.com/capabilities/quantumblack/our-insights/the-state-of-ai-in-2023-generative-ais-breakout-year

Cobbe, J., & Singh, J. (2021). Artificial intelligence as a service: Legal responsibilities, liabilities, and policy challenges. *Computer Law & Security Review*, 42, 105573. https://doi.org/10.1016/j.clsr.2021.105573

Cocca, G. (2023, May 15). How to write clean code – tips and best practices (full handbook). *Freecodecamp.org*. www.freecodecamp.org/news/how-to-write-clean-code/

Cvjetković, M., Vasiljević, M., Cvjetković, M., & Josimović, M. (2021). Impact of quality on improvement of business performance and customer satisfaction. *Journal of Engineering Management and Competitiveness*, *11*(1), 20–28. https://doi.org/10.5937/jemc2101020C

Dahler-Larsen, P. (2019). *Quality*. Springer International Publishing. https://doi.org/10.1007/978-3-030-10392-7

Dastin, J. (2018, October 11). Amazon scraps secret AI recruiting tool that showed bias against women. *Reuters*. www.reuters.com/article/us-amazon-com-jobs-automation-insight-idUSKCN1MK08G

Delipetrev, B., Tsinaraki, C., & Kostic, U. (2020). *Historical evolution of artificial intelligence*. European Commission Joint Research Centre. https://doi.org/10.2760/801580

Deniz, B. K., Gnanasambandam, C., Harrysson, M., Hussin, A., & Srivastava, S. (2023, June 27). *Unleashing developer productivity with generative AI*. McKinsey & Company. www.mckinsey.com/capabilities/mckinsey-digital/our-insights/unleashing-developer-productivity-with-generative-ai

Desjardins, J. (2017, February 8). How many millions of lines of code does it take? *Visual Capitalist*. www.visualcapitalist.com/millions-lines-of-code/

Dumitru, D., & Halpern, D. F. (2023). Critical thinking: Creating job-proof skills for the future of work. *Journal of Intelligence*, *11*(10), 194. https://doi.org/10.3390/jintelligence11100194

Evans, J. R., & Lindsay, W. M. (2013). *Managing for quality and performance excellence* (9th revised ed.). South-Western College Publishing.

Feldman, M. S., & Pentland, B. T. (2003). Reconceptualizing organizational routines as a source of flexibility and change. *Administrative Science Quarterly*, *48*(1), 94–118. https://doi.org/10.2307/3556620

Foster, T., & Gardner, J. (2022). *Managing quality* (7th ed.). Wiley. www.perlego.com/book/3884524/managing-quality-integrating-the-supply-chain-pdf

Fregin, M.-C., Levels, M., de Grip, A., Montizaan, R., & Kensbock, J. (2020, September 3). *Künstliche Intelligenz: Ein sozialpartnerschaftliches Forschungsprojekt untersucht die neue Arbeitswelt*. IBM Deutschland GmbH Und Vereinte Dienstleistungsgewerkschaft Ver.Di. ROA External Reports.

Garvin, D. A. (1987). Competing on the Eight Dimensions of Quality. *Harvard Business Review*, *65*.

Gershgorn, D. (2021, June 29). GitHub and OpenAI launch a new AI tool that generates its own code. *The Verge*. www.theverge.com/2021/6/29/22555777/github-openai-ai-tool-autocomplete-code

Ghaffar Nia, N., Kaplanoglu, E., & Nasab, A. (2023). Evaluation of artificial intelligence techniques in disease diagnosis and prediction. *Discover Artificial Intelligence*, *3*(1), 5. https://doi.org/10.1007/s44163-023-00049-5

Ghosh, S. (2023). *Building low latency applications with C++: Develop a complete low latency trading ecosystem from scratch using modern C++*. Packt Publishing. https://books.google.de/books?id=KiXMEAAAQBAJ

Gieselmann, H., & Trinkwalder, A. (2023). Trügerische Präzision: Wie Benchmarks die Leistung großer Sprachmodelle messen und vergleichen. *C't-Magazin*, *23*(21), 20–23. www.heise.de/select/ct/2023/21/2320813482746375173

Hannay, J. E., Dybå, T., Arisholm, E., & Sjøberg, D. I. K. (2009). The effectiveness of pair programming: A meta-analysis. *Information and Software Technology*, *51*(7), 1110–1122. https://doi.org/10.1016/j.infsof.2009.02.001

Holzinger, A., Saranti, A., Angerschmid, A., Finzel, B., Schmid, U., & Mueller, H. (2023). Toward human-level concept learning: Pattern benchmarking for AI algorithms. *Patterns*, *4*(8), 100788. https://doi.org/10.1016/j.patter.2023.100788

International Organization for Standardization. (2023). *ISO 9000 family*. www.iso.org/iso/quality_management

ISO/IEC. (2011). *ISO/IEC 25010:2011: Systems and software engineering – systems and software quality requirements and evaluation (SQuaRE) – system and software quality models*. www.iso.org/standard/35733.html

Kochhar, P. S., Lo, D., Lawall, J., & Nagappan, N. (2017). Code coverage and postrelease defects: A large-scale study on open source projects. *IEEE Transactions on Reliability*, *66*(4), 1213–1228. https://doi.org/10.1109/TR.2017.2727062

Krasner, H. (2022). *Cost of poor software quality in the U.S.: A 2022 report*. www.it-cisq.org/the-cost-of-poor-quality-software-in-the-us-a-2022-report

Kumar, P. (2016). Quality of quality definitions: An analysis. *International Journal of Scientific Engineering and Technology*, *5*, 142–148. https://doi.org/10.17950/ijset/v5s3/304

Lins, S., Pandl, K. D., Teigeler, H., Thiebes, S., Bayer, C., & Sunyaev, A. (2021). Artificial intelligence as a service. *Business & Information Systems Engineering*, *63*(4), 441–456. https://doi.org/10.1007/s12599-021-00708-w

Liu, B. (2017). Lifelong machine learning: A paradigm for continuous learning. *Frontiers of Computer Science*, *11*(3), 359–361. https://doi.org/10.1007/s11704-016-6903-6

Luthra, S., Garg, D., Agarwal, A., & Mangla, S. K. (2020). *Total quality management (TQM): Principles, methods, and applications*. CRC Press. https://books.google.de/books?id=3K73DwAAQBAJ

Martin, R. C. (2008). *Clean code: A handbook of agile software craftsmanship*. Pearson Education. https://books.google.de/books?id=_i6bDeoCQzsC

Naveed, H., Khan, A. U., Qiu, S., Saqib, M., Anwar, S., Usman, M., Akhtar, N., Barnes, N., & Mian, A. (2023). *A comprehensive overview of large language models*. http://arxiv.org/abs/2307.06435

Ott, S., Barbosa-Silva, A., Blagec, K., Brauner, J., & Samwald, M. (2022). Mapping global dynamics of benchmark creation and saturation in artificial intelligence. *Nature Communications*, *13*(1), 6793. https://doi.org/10.1038/s41467-022-34591-0

Owen, D. B. (2020). *Beating your competition through quality*. CRC Press. https://doi.org/10.4324/9781003066187

Peinl, R., & Wirth, J. (2023). *Evaluation of medium-large language models at zero-shot closed book generative question answering. arXiv Preprint*. arXiv:2305.11991.

Phung, T., Pădurean, V.-A., Cambronero, J., Gulwani, S., Kohn, T., Majumdar, R., Singla, A., & Soares, G. (2023). *Generative AI for programming education: Benchmarking ChatGPT, GPT-4, and human tutors*. http://arxiv.org/abs/2306.17156

Prause, C. R., Werner, J., Hornig, K., Bosecker, S., & Kuhrmann, M. (2017). *Is 100% test coverage a reasonable requirement? Lessons learned from a space software project* (pp. 351–367). Springer. https://doi.org/10.1007/978-3-319-69926-4_25

Reddi, V. J., Kanter, D., Mattson, P., Duke, J., Nguyen, T., Chukka, R., Shiring, K., Tan, K.-S., Charlebois, M., Chou, W., El-Khamy, M., Hong, J., St. John, T., Trinh, C., Buch, M., Mazumder, M., Markovic, R., Atta, T., Cakir, F., . . . Zika, D. (2020). *MLPerf mobile inference benchmark. arXiv Preprint*. arXiv:2012.02328.

Rezabakhsh, B., Bornemann, D., Hansen, U., & Schrader, U. (2006). Consumer power: A comparison of the old economy and the internet economy. *Journal of Consumer Policy*, *29*(1), 3–36. https://doi.org/10.1007/s10603-005-3307-7

Riggins, J., & Gregory, J. (2019, February 10). Who is in charge of quality in software development. *Infoq.com*. www.infoq.com/articles/who-quality-software-development/

Roger, F., Greenblatt, R., Nadeau, M., Shlegeris, B., & Thomas, N. (2023). *Benchmarks for detecting measurement tampering*. http://arxiv.org/abs/2308.15605

Sainz, O., Campos, J. A., García-Ferrero, I., Etxaniz, J., de Lacalle, O. L., & Agirre, E. (2023). *NLP evaluation in trouble: On the need to measure LLM data contamination for each benchmark. arXiv Preprint*. arXiv:2310.18018.

Samuelson, P. (2023). Generative AI meets copyright. *Science*, *381*(6654), 158–161. https://doi.org/10.1126/science.adi0656

Saranya, A., & Subhashini, R. (2023). A systematic review of explainable artificial intelligence models and applications: Recent developments and future trends. *Decision Analytics Journal*, 7, 100230. https://doi.org/10.1016/j.dajour.2023.100230

Sargut, G., & McGrath, R. (2011, September 1). Learning to live with complexity. *Harvard Business Review*, 68–76.

Singh, A., Sharma, S., Kumar, S. R., & Yadav, S. A. (2016). Overview of PaaS and SaaS and its application in cloud computing. *2016 International Conference on Innovation and Challenges in Cyber Security (ICICCS-INBUSH)*, 172–176. https://doi.org/10.1109/ICICCS.2016.7542322

Smith, G. (1993). The meaning of quality. *Total Quality Management*, 4, 235–244. https://doi.org/10.1080/09544129300000038

Stilgoe, J. (2021). How can we know a self-driving car is safe? *Ethics and Information Technology*, *23*(4), 635–647. https://doi.org/10.1007/s10676-021-09602-1

Uddin, A., & Anand, A. (2019). Importance of software testing in the process of software development. *International Journal for Scientific Research and Development*, 2321–2613.

Verhoef, P. C., Broekhuizen, T., Bart, Y., Bhattacharya, A., Qi Dong, J., Fabian, N., & Haenlein, M. (2021). Digital transformation: A multidisciplinary reflection and research agenda. *Journal of Business Research*, 122, 889–901. https://doi.org/10.1016/j.jbusres.2019.09.022

Wang, A., Singh, A., Michael, J., Hill, F., Levy, O., & Bowman, S. R. (2018). *GLUE: A multi-task benchmark and analysis platform for natural language understanding. arXiv Preprint*. arXiv:1804.07461.

Winfield, A. F. T., & Studley, M. (2023). *On the relationship between benchmarking, standards and certification in robotics and AI*. https://arxiv.org/abs/2309.12139

Zaremba, W., Brockman, G., & OpenAI. (2021, August 10). *OpenAI codex*. https://openai.com/blog/openai-codex

Zayour, I., & Hamdar, A. (2016). A qualitative study on debugging under an enterprise IDE. *Information and Software Technology*, *70*, 130–139. https://doi.org/10.1016/j.infsof.2015.10.010

Chapter 10

Copyright Aspects of Generative AI in Germany and the EU

An Overview

Stefan Ellenberg

Introduction

The advent of generative artificial intelligence (AI) has ushered in a new era of media production, transforming the landscape of creative industries across the globe. This powerful technology, capable of creating novel content based on existing data, has revolutionized the way artists and creators approach their work. Skilful use of generative AI enables creators to iterate faster, from the early stages of brainstorming to the high-stakes development phases. This acceleration in the creative process has profound implications for animation, film, video games, design, and more.

The impact of generative AI is though not limited to professionals alone; it has democratized creativity and lowered barriers to entry for individuals across various artistic domains. Previously, fields such as animation, TV, film, and design may have been accessible primarily to well-established studios. However, with generative AI, any person, with the right skills and knowledge, can now effectively function as a studio head, empowered to produce innovative and high-quality content independently.

Generative AI is being used in various forms of media production, including news, publishing, film, gaming, and music (an overview can be found at https://internationalsales.lexisnexis.com/news-and-events/everything-to-know-about-generative-ai-in-media-production). For example, generative AI is being used to create new forms of news content, such as audio and video summaries of news articles (www.ap.org/discover/artificial-intelligence). In the film industry, generative AI contributes to realistic visual effects, virtual sets (https://deadline.com/2023/05/ai-unreal-engine-technology-disruptors-1235364383/), and even aids in dubbing films for foreign screenings (www.flawlessai.com/). It has the capability to create younger-looking versions of lead actors (https://fortune.com/2023/02/14/tech-forward-everyday-ai-hollywood-movies/), expanding creative possibilities and storytelling potential. Gaming experiences are enriched through generative AI, which facilitates personalized gameplay and AI-powered opponents that adapt to players' strategies (www.aiacceleratorinstitute.

DOI: 10.4324/9781003453901-13

com/artificial-intelligence-in-gaming-revolutionizing-the-player-experience/). Music creators are leveraging generative AI to generate songs and compose soundtracks, opening doors to a whole new world of musical expressions (www.linkedin.com/feed/update/urn:li:activity:7056704609683132416/).

However, these advancements raise new legal questions, particularly in the field of copyright law. The use of AI to create content raises important questions such as the lawfulness of AI training, the copyright protection for AI autonomous and co-created output, and the infringement of third-party rights associated with AI-created content.

Generative AI and Copyright Law

Text and Data Mining

A contentious global issue related to generative AI is whether the use of copyrighted materials for training purposes requires consent from the original authors and other rights holders. Particularly in the United States and the United Kingdom, numerous ongoing legal proceedings find authors and other rights owners opposing the use of their works for training of generative AI.

On January 13, 2023, several artists initiated a presumptive class-action lawsuit in California, contending that their copyrights were violated during the training of AI image programs, notably Midjourney and Stable Diffusion (https://stablediffusionlitigation.com/pdf/00201/1-1-stable-diffusion-complaint.pdf).

In a similar vein, on February 3, 2023, Getty Images lodged a lawsuit in Delaware, alleging, "Stability AI has replicated at least 12 million copyrighted images from Getty Images' platforms... in its endeavor to train the Stable Diffusion model" (https://copyrightlately.com/pdfviewer/getty-images-v-stability-ai-complaint/?auto_viewer= true#page =&zoom=auto&pagemode=none).

The authors contend, with reasonable justification, that generative AI can produce images, texts, and sounds primarily because it was trained with their copyrighted works. Only through such training can the AI compete with the authors' original creations. Consequently, they argue, they should be adequately compensated for the use of their works. However, whether the claims of authors and rights holders indeed hold water significantly depends on the jurisdiction in which these claims are made. For instance, the current legal landscape in the Anglo-American legal domain distinctly differs from that within the European Union (EU) and Germany. It might sound paradoxical, but currently in the copyright-friendly European and German legal environment, such lawsuits would likely face challenges in gaining traction. The reason for this lies though not within the actual training of AI and the question if this constitutes a copyright-relevant use, but in the statutory limitations of the respective copyright laws.

Copyright Relevance of AI Training in Germany

Given that the German copyright law aims to facilitate the advancement of art and culture, it is permitted to read, listen to, or view a work and extract information from it for one's own creation (Bundestagsdrucksache IV/270, S. 28, 63). Hence, copyright does not protect the information contained within a work but rather its specific, syntactic embodiment in the work (Maamar, 2023a, pp. 481–482). This is the same for humans and large language models. Analysing a work through AI – in principle – does not differ from a corresponding human engagement with the work. The actual training of AI, that is, the automated analysis of training data, is therefore possible without permission. However, to analyse the data, it must be saved beforehand. The associated duplication affects the reproduction right of the authors according to §16 of the German Copyright Act (UrhG). To legally train AI models, copyrighted data must therefore either be licensed or the AI provider must be able to rely on a limitation of copyright. And in this very respect, the legal landscape in the EU and Germany differs significantly from that in the Anglo-American and many other legal systems, which do not have a specific statutory exemption for text and data mining.

Exemption for Text and Data Mining in the EU and Germany

Before the passage of the Digital Single Market (DSM) Directive in 2019 (Directive [EU] 2019/790 of the European Parliament and of the Council on copyright and related rights in the Digital Single Market and amending Directives 96/9/EC and 2001/29/EC), public attention was primarily focused on the liability of platforms through upload filters (Raue, 2021, pp. 793–794). Amidst this, another significant yet largely unnoticed change was introduced: the expansion of copyright limitations to include "Text and Data Mining" for commercial purposes, as set out in Article 4 of the DSM Directive.

According to Article 4 No. 1 of the DSM Directive, all EU member states shall provide for an exception or limitation to certain exclusive rights of authors for reproductions and extractions of lawfully accessible works for the purposes of text and data mining. In Germany, this provision was implemented in §44b of the UrhG, which mirrors the respective requirements of the DSM Directive. The legal situation in the EU and Germany is therefore the same.

Section 44b of the UrhG defines text and data mining as "*any automated analytical technique aimed at analysing text and data in digital form in order to generate information which includes but is not limited to patterns, trends and correlations*". The reason for the limitations set out in §44b of the UrhG is the fact that

> *Text and Data Mining methods are not only of great significance in the context of scientific research, but they are also widely used by both private and*

> *public institutions to analyse vast amounts of data across various aspects of life and for diverse purposes, including government services, complex business decisions, and the development of new applications or technologies.*
> (No. 18 of the recitals of the DSM-Directive as referenced by Bundestagsdrucksache 19/27426, 87)

Thus, the question arises whether the *training* of generative AI serves the purpose of analysing data and can therefore be subsumed under the term "*text and data mining*" as set out in §44b of the UrhG.

First, it must be noted in this respect that "*data*" in the sense of §44b of the UrhG refers to data of any categorization. Therefore, the term *data* particularly encompasses sounds, images, and films. It also covers the digitization of analogue information for training purposes (Raue, ZUM 21, 793, 795). Furthermore, to limit the rights of the authors, generative AI training would need to aim at the extraction of information from the trained content. If this is the case, depends on the functionality of generative AI. For example, generative AI models are based on machine learning (Pesch & Böhme, 2023, pp. 997–998). Machine learning, in turn, focuses on generating knowledge from experience, that is from existing data, rather than from predefined rules. For this very purpose, the training data are provided to the AI. Using the so-called neural networks, the AI then analyses the provided information and recognizes patterns and similarities in the data. It can then subsequently derive abstract rules from them and incorporate them into the model's parameters (Klawonn, 2023, p. 36; Pesch & Böhme, 2023, pp. 997, 1006). Hence, the AI doesn't have abstract rules set for it, but instead learns the rules on its own from analysing the data. By applying these rules, the AI is then capable of accomplishing tasks, such as generating content, which would otherwise be unsolvable. Since the training of generative AI serves the derivation of abstract rules and their translation into the model's parameters, the training of generative AI models falls, by the prevailing opinion, under the definition of "*text and data mining*" as per §44b of the UrhG (Maamar, 2023a, pp. 481–482; Pesch & Böhme, 2023, pp. 997, 1006; Schack, 2021, 904). However, this opinion is not uncontested. It is argued that the characteristics and patterns recognized by AI are not output as information, but rather remain within the system and are difficult to interpret even for experts (Pukas, 2023, p. 614). It will have to be seen how the courts will eventually decide on this matter.

In any case, for the training of generative AI to be permitted, the conditions of Article 2 No. 2 of the DSM Directive, §44b of the UrhG, respectively, must be met, which correspondent with each other as stated above. These conditions are primarily:

Legitimate Accessibility of Training Material

First, any "*Text and Data Mining*" is restricted to works that are "*legally accessible*". A work – including databases and computer programs and other protected objects of the UrhG, such as related rights of performers and phonogram producers in sound recordings (Bundestagsdrucksache 19/27426, 88.) – is legally accessible if the user can access it without violating the law (Maamar, 2023a, pp. 481–482). However, it is not required that the work itself was made available legally (Dreier/Schulze/Dreier, UrhG §44b Rn. 8). Therefore, in principle, all freely accessible works on the internet can be used for AI training purposes (Raue, 2021, pp. 794, 796). This includes works to which the user has access due to contractual agreements (Raue, 2021, pp. 793, 800).

Obligation to Delete After Training Completion

The second requirement that must be met is the obligation to delete the stored training data once the training has been completed. Any storage beyond that is not permitted, which also includes future training purposes. A longer storage can though be permitted in case of §60d of the UrhG for scientific research. Another way to circumvent the deletion obligation is licensing agreements. However, due to the sheer volume of data, this may be challenging to achieve. Therefore, there's an ongoing discussion about whether collecting societies can grant such licences as so-called extended collective licences, according to §§51 ff of the VVG, which is based on Articles 8 and 12 of the DSM Directive (on this Pukas, 2023, p. 614 ff.).

No Opt-out for Rights Holders

Last, the rights holders must not have objected to the use of their content. This is referred to as the "*opt-out*". The need for rights holders to object to the use represents a reversal from the principle usually applicable in copyright law, where the user must actively seek the consent of the rights holder. In this case, the rights holder must take the initiative to prevent usage. It's still unclear how the rights holder must declare the *opt-out*. The wording of the law only mentions that the objection must be made in a "machine-readable form". While this can be included in the imprint and terms and conditions, it must still be in machine-readable form. Objections in natural language are, in contrast, irrelevant. The requirement of the opt-out to be declared in a machine-readable form is to create legal certainty for AI providers (Bundestagsdrucksache 19/27426, 89). The AI provider, in turn, has the burden of proof that the rights holder did not declare an opt-out before the content was copied. The opt-out only takes effect from the point at which it is declared. Copies made for training purposes before that are therefore not covered. Notably, an *opt-out* for a website must not result in it being treated differently in other uses without object justification, for example when displayed as a search engine results (Recital 18 subparagraph 2 sentence 3 DSM-RL; Bundestagsdrucksache 19/27426, 89).

Geographical Scope of Protection

Whether the training of AI according to §44b of the UrhG or another limitation provision is permissible is also determined by the location where the necessary duplication into the AI corpus took place. If this was outside Germany or the EU, the AI provider cannot rely on the limitation provisions of the UrhG or the DSM Directive (Maamar, 2023b, pp. 481, 484).

Comparative Perspective: Legal Situation in the United States

While the primary focus of this chapter is on Germany and the EU, a brief overview of the legal situation in the United States is provided for a comparative understanding. This allows us to appreciate the global diversity in legal approaches to generative AI. As mentioned earlier, the United States lacks explicit legal regulation regarding the training of generative AI. The legal disputes there mainly revolve around whether the duplications necessary for training, of copyrighted content, could be considered "*fair use*" under 17 U.S.C. § 107.

The following four prerequisites would have to be met:

1. *The purpose and character of the use, including whether such use is of a commercial nature or is for nonprofit educational purposes.*
2. *The nature of the copyrighted work.*
3. *The amount and substantiality of the portion used in relation to the copyrighted work as a whole.*
4. *The effect of the use upon the potential market for or value of the copyrighted work.*

Proponents of "*fair use*" argue, among other things, that the results of generative AI are of a transformative nature and do not merely reproduce the utilized works (www.uspto.gov/sites/default/files/documents/OpenAI_RFC-84-FR-58141.pdf#page=5). Additionally, it is argued that the copies are only stored in the AI corpus and are not made available to the public as such.

Opponents, on the other hand, argue that the negative impacts on the market and the value of copyrighted works can be quite significant. Examples include the AI-generated song "Heart on My Sleeve", which mimics Drake and The Weeknd, and was streamed millions of times in April 2023 (www.theverge.com/2023/5/1/23703087/ai-drake-the-weeknd-music-copyright-legal-battle-right-of-publicity). The same applies to book authors under whose names AI-generated books are sold on Amazon but were not actually written by these authors (https://edition.cnn.com/2023/08/10/tech/ai-generated-books-amazon/index.html).

It remains to be seen, with great anticipation, how the US courts will decide in the aforementioned cases.

Copyright Protection for AI-generated content in the EU and Germany

Besides the question of the legitimacy of AI training, the copyrightability of the results (i.e. the output) is the most discussed aspect in connection with generative AI models at the moment, most likely because the creative industry is ultimately based on licensing agreements that grant rights to the created works. However, if there are no rights in the output, then the creator cannot grant those to its contractual partner and, moreover, the created content is not protected against third-party use. In both cases, this would have immense implications for the functioning of the creative industry, as it is built on the grant of rights.

Author – Human and Machine?

Who should be deemed as the author? Human and machine? Or only humans? Decisive for this is how the term "*work*" is defined in European and German copyright laws, as the author is deemed as the creator of a work.

According to the rulings of the Court of Justice of the European Union (CJEU), the concept of "*work*" in copyright encompasses two conditions. First, it entails an original subject matter which is the author's own intellectual creation and, second, it requires the expression of that creation (judgment of 12 September 2019, *Cofemel*, C-683/17, EU:C:2019:721, paras 29 and 32 and the case-law cited). In that regard, it must be borne in mind that, if creating something is fully dictated by technical rules or constraints, leaving no room for creativity, it cannot be considered original. Therefore, it won't be protected by copyright (judgment of 12 September 2019, *Cofemel*, C-683/17, EU:C:2019:721, para 31 and the case-law cited). This is an autonomous concept of EU law that needs to be interpreted and applied uniformly throughout the EU. In essence, this definition aligns with the legal position in Germany.

Although there hasn't been an explicit determination by the CJEU on this yet, the requirement for an "*own intellectual creation*" of the author suggests that a "*work*" in terms of copyright will always require human creation. In addition, the wording of the UrhG, which speaks of a "*personal intellectual*" creation, emphasizes this even more. As the output of a generative AI is ultimately not "*created*" by a human, it cannot be a "*work*" in the sense of §§2 and 7 of the UrhG, and it is therefore not protected as such (Lauber-Rönsberg, p. 4.).

Having said this, it is clear that AI will – at least for the moment and from a legal standpoint – not generate content on its own. Hence, the output of generative AI is always subject to the instructions given by the human to the AI, that is the prompt.

The relevant aspect for the assignment of copyright on AI-generated content is therefore the level of influence the human has on the outcome. If the human specifically dictates the outcome through the prompt, leaving no room for AI interpretation, this would induce for a human authorship as the AI would then be considered as a tool used by the author. However, if the AI retains a level of creative freedom, this argues against the human's authorship. On the other hand, it is also argued that it should be sufficient for the qualification as copyright-protected work, if the creative choice made in the concept stage is expressed in the final AI-assisted output (Hartmann/Allan/Hugenholtz/Quintais/Gervais, p. 117). Therefore, whether AI-generated works can be copyright-protected as works can't be generalized but is based on human influence in specific cases (Klawonn, p. 94; Marmaar, p. 191).

Another exception to the principle that works autonomously generated by an AI aren't subject to copyright are in Germany, the so-called "*neighboring rights*". This primarily concerns the neighboring rights of film producers, as per §94 of the UrhG, and of record producers, according to §85 of the UrhG. Whether a similar protection also exists for databases is a matter of debate (see von Welser, 2023, pp. 57, 59). Unlike copyright, the neighboring rights don't hinge on a creative performance but protect the investment made in a particular performance result (see CJEU, 2019, p. 738 Rn. 30 – Metall auf Metall III; Recitals 10 InfoSoc-Directive) (e.g. music or film recording). In this respect, generative AI providers or their users do not differ from music labels or film producers. Reproduction, distribution, and public accessibility of AI-generated music or AI-generated films are therefore only permissible in Germany with the consent of the producer (Maamar, 2023b, pp. 481, 490; Klawonn, p. 95f.). This can either be an AI-model provider or the user who prompts and typically pays for the generated content.

Through Article 2 lit c. and d. of the InfoSoc Directive, the rights of the record producer and the film producer to reproduce the content are also protected within the EU. Beyond Germany and the EU, comparable protection only exists in jurisdictions that recognize similar investment protection (see Article 3 of the Treaty on the Harmonization of Copyright and Neighboring Rights "WCT").

Comparative Perspective: Legal Situation in the United States

The legal situation in Germany and the EU as set out above is consistent with the current legal situation in the United States. The US Copyright Office explicitly protects only works that have been created by humans. This is because US copyright law only protects "*the fruits of intellectual labor*" that "*are founded in the creative powers of the mind*". Copyright law is confined to "*original intellectual conceptions of the author*" (as referenced in the Trade-Mark Cases, 100 U.S. 82, 94 (1879)). As such, the US Copyright Office will refuse to register a claim if it determines that a human being did not create the work (Compendium of the US Copyright Office Practices,

Third Edition, No. 307). In line with this, the US Copyright Office granted only partial protection for a graphic novel titled *Zarya of the Dawn*, which artist Kris Kashtanova had partially created with the assistance of Midjourney. On February 21, 2023, the Copyright Office determined that the images were not copyrightable, deciding that Midjourney, rather than Kashtanova, authored the "*visual material*" (https://fingfx.thomsonreuters.com/gfx/legaldocs/klpygnkyrpg/AI%20COPYRIGHT%20decision.pdf). Employing the same reasoning in *Stephen Thaler* (*Stephen Thaler v Shira Perlmutter and The United States Copyright Office* [1:22-cv-01564]), a request for summary judgment over the refusal of his application for a work generated entirely using generative AI technology was granted to the US Copyright Office (www.courtlistener.com/docket/63356475/23/thaler-v-perlmutter/).

Comparative Perspective: Legal Situation in the United Kingdom and Ex-Commonwealth

In contrast to the approaches in Germany, the EU, and the United States, the legal situation in the United Kingdom regarding copyright protection for AI-generated works presents a unique perspective that highlights the diversity of legal responses to AI-generated works. According to §9(3) of the Copyright, Designs and Patents Act 1988, computer-generated works are also protected by copyright. The author is considered to be the person "*by whom the arrangements necessary for the creation of the work are undertaken*". This is the world's first regulation concerning works generated by a computer, although the rule originally referred to fully automated processes when taking satellite photos. Corresponding regulations are also found in §11(3) of the Hong Kong Copyright Ordinance; §21(f) of the Irish Copyright and Related Rights Act 2000; §5(2a) of the New Zealand Copyright Act; and in §1(h) of the South Africa Copyright Act of 1978. However, whether the Irish rule is consistent with the case law of the CJEU is questionable (see Hartmann/Allan/Hugenholtz/Quintais/Gervais, p. 88; von Welser, 2023, pp. 57, 59). If Irish courts ultimately grant copyright protection to AI-generated output, a referral to the CJEU under Article 267 III of the Treaty on the Functioning of the European Union would likely be necessary.

Infringements

Just as with training, that is on the input side, AI-generated output can lead to conflicts with third-party rights. For example:

Input and Uploads

First, it should be noted that the input or upload of copyrighted content into the AI is only permissible if the user either owns or has acquired the rights for this or if statutory exemptions in favour of the user exist. Accordingly, this also applies to the modification and transformation of content.

One exemption that could be considered here is the right to private copying, according to §53 of the UrhG. This provision allows for individual reproductions by a natural person for private use, provided they are not directly or indirectly intended for commercial purposes. However, even for private use, users will typically want to modify the content, which the exemption does not cover. In any case, the mere act of uploading content for training purposes by the user does not seem comparable to the private copy as envisioned by the exemption. The user does not simply intend to copy, but to utilize the content for transformative purposes. Therefore, it is unlikely that an upload will be considered as a private copy according to §53 of the UrhG.

Reproduction of Existing Works

Furthermore, it is possible that copyrighted works of third parties might be included in the output. It is important to note that any use constitutes an infringement of rights from the outset if the reproduced works should not have been used for training the AI. Whether the user is aware of this is not relevant. The claim for injunction according to §97 Abs.1 S.1 of the UrhG is independent of negligence (see only BeckOK UrhR/Reber, UrhG, §97 para. 91). As for the question of the permissibility of training, it also depends on the location where the AI training took place.

Moreover, when a work is reproduced in its entirety, an infringement is generally presumed due to the associated duplication. An exception might only be conceivable if the incorporated work is entirely subsumed in the output – for instance, due to varying size ratios – and is no longer recognizable as such.

In cases where the works of third parties are partially incorporated into the output, the degree of difference between the output and the original work becomes critical. Without sufficient differentiation, the publication and exploitation of the output, without the consent of the original copyright holder, are fundamentally impermissible under §23 Abs.1 S. 1 of the UrhG.

Whether there is sufficient differentiation depends on whether the incorporated elements are subsumed into the new work and do not characterize the output in such a way that they constitute the essence of the output (Wandtke/Bullinger/Bullinger, UrhG, §23 para. 36). The CJEU further specifies this, referring to the rights of record producers in Article 2 of the InfoSoc-Directive (Directive 2001/29/EC of the European Parliament and of the Council of 22.5.2001 on the harmonization of certain aspects of copyright and related rights in the information society (OJ 2001 L 167, 10)). The court posits that the only permissible adaptation is one that has been changed in a manner unrecognizable when the work is consumed, and this without the consent of the entitled party (CJEU, 2019, p. 929 para. 39 – Metal on Metal III).

Therefore, whether this condition is met cannot be assessed generically, but always depends on the circumstances of the individual case. However, the more identifiable the incorporated part in the output, the more likely the consent of the rights holder will be necessary. Exceptions might only arise if the recognizable adaptation is done for purposes of parody, satire, or pastiche.

It should also be noted that technically induced modifications in the context of AI training – such as anomalies in diffusion models – do not require the consent of the rights holder, according to §23 Abs. 3 of the UrhG, as they are purely technically conditioned.

Contents "in the Style of"

From a copyright perspective, generating content in the style of a particular artist is unproblematic. This is because style, per se, is not protected; only the specific design of a work is (see §2 of the UrhG). The artist's underlying style remains in the public domain to ensure that an artist does not monopolize a chosen cultural technique through copyright, thereby restricting the artistic freedom of others. Just as a human can, so too can an AI imitate an artist's style. However, imitating can become legally problematic if the output is published or disseminated as a work of the imitated artist.

Deepfakes

In the case of so-called *Deepfakes*, which are deceptively real-looking video, image, and audio recordings, there is typically an infringement on the personal rights of the individuals depicted. Additionally, there may also be claims arising from data protection law due to the processing of personal data. Exceptions might only arise in cases of clearly satirical representations when the protection of the personal rights of the affected individual does not predominate (in this respect see Kumkar & Rapp, 2022, p. 199 ff.).

Future Research Agenda

The discourse in this chapter has unveiled many unresolved questions, necessitating further exploration. In the rapidly evolving realm of generative AI, legal frameworks in Germany, the EU, and beyond face unprecedented challenges. In the following, a few research questions are raised, aiming to dissect these complexities and offer insights into the future trajectory of copyright law in relation to generative AI.

RQ1: Legal protection for autonomously created AI-generated content?
As explained, AI-generated content, with few exceptions, does not fall under the existing copyright legislation. The reason for this is the prevailing opinion that only humans are capable of an own intellectual creation. Given the influence that generative AI has and will have on the processes in the

creative industry, the question arises whether the current understanding of authorship is still contemporary. Considering the influence users can exert through prompts, for instance, on the creation of an image in Midjourney, one must question how this differs from a photographer's role. In both scenarios, aspects such as the motif, frame, and lighting, can be chosen with full creative discretion. In both cases, a technical device, not a human, creates the image itself. Therefore, there's a need for in-depth engagement with the question of why copyright is granted and whether this justification remains valid exclusively concerning humans (in this respect see Scarnell, 2022, *Journal of Intellectual Property Law & Practice**, *17*, 727–740). A follow-up question would then be, to whom any eventual copyright should be attributed to (in this respect see Hartmann et al., 2020, p. 117).

RQ2: Evidentiary issues in relation to AI-generated content?

As we have observed, the protection under copyright law currently still depends on content being entirely or predominantly created by humans. However, often today, AI-generated content can no longer be distinguished from human-created ones. Given this backdrop, the question arises as to how and by whom authorship can and must be proven in judicial or administrative proceedings (in this respect see Henke, 2023, pp. 608–614).

RQ3: Mandatory labelling for AI-generated content?

Given the quality that AI-generated content already achieves today, the question arises whether there should be a mandatory labelling for such content. This is for two primary reasons: first, there is an increasing risk of individual and public deception due to the spread of realistically appearing yet ultimately false content created by generative AI. Second, the costs associated with producing AI-generated content are usually significantly lower than those incurred by human endeavours. Consequently, AI-generated content leads to a competitive environment substantially skewed in favour of AI. If we wish to preserve human creativity in the future, the prerequisites for this, including financial compensations, must be legally safeguarded. One possible measure could be mandatory labelling for AI-generated content. However, whether this approach is indeed necessary and promises legal and financial success requires further investigation (in this respect see Gerecke, 2023, pp. 381–384; Hinderks, 2022, pp. 110–119).

References

Dreier/Schulze. (2022). *UrhG, 7. auflage*. https://www.beck-shop.de/dreier-schulze-urheberrechtsgesetz-urhg/product/32549155

Gerecke, M. (2023). Social Media und Recht: Einige urheberrechtliche Gedanken zu generativen KI-Modellen. *Gewerblicher Rechtsschutz und Urheberrecht, Praxis im Immaterialgüter- und Wettbewerbsrecht (GRUR-Prax)*, S381–S384.

Hannes Henke. (2023). Künstliche Intelligenz als Herausforderung im urheberrechtlichen Verletzungsprozess. *Gewerblicher Rechtsschutz und Urheberrecht (GRUR)*,

S608–S614. https://www.grur.org/uploads/media/GRUR_9_2023_Inhaltsverzeichnis_03.pdf

Hartmann, C., Allan, J. E. M., Hugenholtz, P. B., Quintais, J. P., & Gervais, D. (2020). *Trends and developments in artificial intelligence challenges to the intellectual property rights framework*. https://www.ivir.nl/publicaties/download/Trends_and_Developments_in_Artificial_Intelligence-1.pdf

Hinderks, T. (2022). Die Kennzeichnungspflicht von Deepfakes. *Zeitschrift für Urheber- und Medienrecht (ZUM)*, S110–S119. https://beck-online.beck.de/Dokument?vpath=bibdata%2Fzeits%2Fzum%2F2022%2Fcont%2Fzum.2022.110.1.htm&pos=12

Klawonn, T. (2023). *Künstliche Intelligenz, Musik und das Urheberrecht*. Mohr Siebeck.

Kumkar, L. K., & Rapp, J. P. (2022). Deepfakes: Eine Herausforderung für die Rechtsordnung. *Zeitschrift für Digitalisierung und. Recht (ZfDR)*, S199–S228.

Maamar, N. (2023a). *Computer als Schöpfer*. https://www.mohrsiebeck.com/buch/computer-als-schoepfer-9783161608728?no_cache=1

Maamar, N. (2023b). Urheberrechtliche Fragen beim Einsatz von generativen KI-Systemen. *Zeitschrift für Urheber- und Medienrecht (ZUM)*, S481–S491.

Pesch, P. J., & Böhme, R. (2023). Artpocalypse now? – Generative KI und die Vervielfältigung von Trainingsbildern. *Gewerblicher Rechtsschutz und Urheberrecht (GRUR)*, S997–S1007.

Raue, B. (2021). Die Freistellung von Datenanalysen durch den neuen Text und Data Mining-Schranken (§§44b, 60d UrhG). *Zeitschrift für Urheber- und Medienrecht (ZUM)*, S793–S802.

Schack, H. (2021). Schutzgegenstand, "Ausnahmen oder Beschränkungen" des Urheberrechts. *Gewerblicher Rechtsschutz und Urheberrecht (GRUR)*, 904–909.

Scarnell, B. (2022, September). When Irish AIs are smiling: Could Ireland's legislative approach be a model for resolving AI authorship for EU member states? *Journal of Intellectual Property Law & Practice*, 17(9), 727–740.

von Welser, M. (2023). ChatGPT und Urheberrecht, Gewerblicher Rechtsschutz und Urheberrecht. *Praxis im Immaterialgüter- und Wettbewerbsrecht (GRUR-Prax)*, S57–S59.

Wandtke/Bullinger. (2022). *UrhG, Kommentar, 6. Auflage*. https://www.beck-shop.de/praxiskommentar-urheberrecht-urhr/product/31607794

Chapter 11

AI Creativity and Human Enhancement

The Identity Link

Kristina Khutsishvili

Introduction

The notion of artificial intelligence (AI) at the time of invention could be viewed as an oxymoron: intelligence, traditionally regarded as an exclusively 'human' attribute, became ascribed to something 'artificial', a product of human ingenuity. In contemporary discourse, we do not reflect on the literal meaning of the AI term. However, the emergence of a newer concept, AI creativity, reintroduces a similar paradox: a quality once deemed exclusive to human cognition, with a potential extension to non-human animals, is now associated with a human-made technological artefact.

The chapter adds insights to the AI creativity discussion, arguing for the relevance of the human enhancement framework. The view on AI creativity as a possible tool for human enhancement, being complimentary to human creativity, is not the common one, as the debate is primarily focused on the dichotomic view on the possibility of a non-human agent (machine, AI system) being seen as creative (Colton et al., 2009; Nikkel, 2023). While the opponents of the possibility of creativity attribution to an AI system may use *augmentation* wording while describing the added value of AI tools in a human-driven creative process, the notion in these cases is used outside of the context of human enhancement debate (Eapen et al., 2023).

Technological advancements have given humans the potential to intervene and modify themselves, starting with their bodies. The ways the modifications can be exercised may be differentiated by categories based on temporality versus permanence: reversible (through wearable robotics), permanent (through non-removable implants), or even inheritable (through the means of genetic engineering). The emergence of these possibilities gives a rise to a range of ethical and legal considerations, with the 'list' constantly updating, in step with the rapid pace of technological developments. The debates are not only stirred by the tangible technological tools under the umbrella of human enhancement but also by the possibilities themselves, with imagination and imaginaries playing a crucial role (Cornips & van Asselt, 2013; Lüthy & Koops, 2013; Nelis & Posthuma, 2013; Svana, 2017).

DOI: 10.4324/9781003453901-14

As with other means of technology, the potential benefits are discussed on par with the possible risks. What makes the human enhancement debate particularly complex is the presence of several aspects: the power of new emerging technologies including AI, the ongoing and rapid technological development, the factor of imagination referring to the corpus of science fiction imaginaries, the underlying ontological questions of what it means to be a human, with the views that may derive from a wide range of standpoints be it a philosophical, biomedical, or religious studies corpora. The ambition of the debate is even sharpened when it concerns the matter of creativity that up until recently was seen as a predominantly human characteristic, with non-human animals sharing this ability with humans (Kaufman & O'Hearn, 2017). The discussion on the (im)possibility of machine and particularly AI creativity may be seen as radical, not only in the context of the debate on the future of work but beyond, as seeking for an ontological answer.

On a conceptual level, we may summarise the concerns on the future of work by the question 'Are we becoming "less" than we were?' – in the context of rapid technological development and the possibility of some of these technologies substituting humans in their professional occupations. When it comes to creativity, the very same question extends its existential significance to the 'highest' sphere of competence, almost spiritual, that is expressed by the top level of Maslow's pyramid of needs. The act of creating relates not only to self-expression and realisation but also to the sense of purpose. The creative process itself may be seen as a way of expressing spiritual concepts, beliefs, or experiences.

The possibility of self-perception as being 'less' than before in the context of artistic expression and consequent self-realisation adds a layer of complexity to an individual's view on AI creativity. The narratives constituting such a reflection are of an identitarian nature, with an AI system being seen not just as a substitute for human ability – which summarises the concerns on the future of work – but as a competitor, a 'machine' that starts to be seen as something 'more' than merely a tool or a technological artefact.

While the practical means and tools that can be viewed under the 'umbrella' of the human enhancement concept have the potential to alleviate inequalities and disabilities, widening the scope of human capabilities, the progress on a collective level is accompanied by societal anxieties, and on an individual – by an individual perception of risk and a view on potential benefit.

Human Enhancement Paradigm

The Context

The human enhancement paradigm involves envisioning possibilities for augmenting human abilities and functions going beyond the boundaries of what was traditionally considered a 'norm'. By human enhancement Juengst

(1997, p. 29) understands 'interventions designed to improve human form or functioning beyond what is necessary to sustain or restore good health'.

According to Almeida and Diogo (2019, p. 184), 'human enhancement can be considered as an improvement of the "limitations" of a "natural version" of the human species with respect to a specific reference in time, and to different environments'. As outlined by Bavelier et al. (2019, p. 204), human enhancement technologies 'could in principle extend capacity beyond the typical range of human experience, posing new ethical challenges at both the individual and collective levels'.

The term 'human enhancement' emerged in the twentieth century. The human enhancement paradigm has been tightly related to the spheres of imagination and imaginaries: at that time, practical tools for human enhancement had not yet been invented, but discussions on possibilities and related moral aspects were present. Scientific and technological progress was envisioned through works of imagination and creative means.

Imagination allowed thinkers, artists, and scientists to explore the potential future scenarios where technological advancements could enhance various aspects of human nature. The expression of imagination and creativity within the context of scientific and technological progress itself but also the envisioned developments turned itself into the pages of fiction books, movies, and other forms of artistic expression. Works of science fiction, for example, often served as a creative space to contemplate the consequences of human enhancement, portraying utopian or dystopian futures shaped by the imaginative exploration of advanced technologies. Such narratives laid the groundwork for ethical considerations and debates on the desirability and implications of pursuing human enhancement technologies. Questions surrounding the potential impact on human identity, equality, autonomy, and societal values are deeply intertwined with the imaginative exploration of what it means to enhance human capabilities.

The debate on human enhancement, which developed throughout the first decade of the twenty-first century, has been polarised through two ideologically opposed viewpoints: transhumanism and bioconservatism. The former sees the current form of the human species as only one stage of human development and emphasises the integration of natural and artificial elements, while the latter emphasises the need to investigate the implications of technological interventions on human dignity and the concept of nature. As summarised by Bostrom (2005, p. 203), transhumanists 'believe that human enhancement technologies should be made widely available, that individuals should have broad discretion over which of these technologies to apply to themselves', while bioconservatives 'are generally opposed to the use of technology to modify human nature', with the main argument of human enhancement technologies having the potential to 'undermine our human dignity'.

As the questions on human enhancement affect the entire 'human condition', interdisciplinary approaches are both suitable and relevant if we aim to understand this complex phenomenon better. However, it is challenging to identify a clear set of technologies explicitly oriented towards human enhancement, given their highly diverging and incomparable characteristics. Innovative brain and neuro-technologies raise ethical, social, and legal issues that intersect with the perspectives of enhancement and neuro-enhancement by means of robotic devices. These issues lead to the question of what makes us human and what is constitutive in our nature as conscious, sentient, and thinking beings. The literature examining these questions is growing, but there is still a lack of systematic, consistent, and concise analysis of the ethical, social, and legal issues and their intersection. The complexity also derives from the variety of the disciplines and methodological standpoints through which the issues may be approached. Medicine, philosophy, engineering, informatics, economics, jurisprudence, and other fields and subfields of human knowledge are all relevant and demanded for the debate, and it is reasonable to expect that such a debate will be informed by different disciplines, with the discussion process being open-ended and hence prepared to include new developments, possibilities – both in their potential benefit and risk dimensions, trends, and various aspects. Yet inter- and transdisciplinarity brings to the forefront the need for translation – of the central concepts, methodologies, and approaches underlying the disciplinary focus. While collaboration and sharing between different disciplines seems to be not only desirable but necessary for tackling complex and ongoing issues, it brings its challenges and the need for the effort to be made to facilitate the successful exchange.

Human enhancement is a broad concept often approached through specific standpoints, such as transhumanism or bioconservatism. These viewpoints may be considered as somehow 'ideological', reflecting the static views on the issue rather than remaining open to subject-related reflection. With societal anxieties surrounding current rapid technological developments, the human enhancement debate, built on fundamental questions, such as what it means to be human, becomes 'ideologically' polarised and a focal point for individual and collective fears related to the future.

In contrast to debates on other emerging technologies, the human enhancement debate is more susceptible to 'existential risks' arguments, as the scenario of further human–machine convergence can be viewed as a logical continuation of human enhancement practices. Science fiction references are intrinsic to the human enhancement debate, also making the link with existential risks argumentation.

The theme of risk in relation to human enhancement technologies can be seen as identity-related and situated on two levels, analogous to the personal and collective 'levels' of identity. Societal anxieties stemming from rapid technological developments are relevant to the collective level of risk perception, while an individual's view on risks and potential benefits relates to an individual level of perception.

The process of collective identity formation is seen through the lens of imagination, playing an important role in human enhancement perception as well. Imagination allows individuals to envision possible futures where human enhancement tools are widely used. People may imagine scenarios where technologies enhance the physical, cognitive, sensory abilities, and the way potential benefits and risks are perceived depends on this 'picture', a consequence of the imagination process.

Societal anxieties and individual fears rely to a significant extent on imaginaries derived from the artefacts of science fiction, reflecting the imagination processes of authors and members of the audience.

While processes on the collective level need to be addressed through an open-ended dialogue, resembling the process of identity formation, individual-level processes are tightly related to the concrete perception of risks and benefits on a personal level. This is not the sole connection between the views on human enhancement, risks, and identities. Technologies aimed at human enhancement may be seen as having the potential to affect one's sense of the self. There is a view that technologies aimed at human enhancement, transforming individuals' physical and intellectual abilities, may be able to alter personal identities. The directions in which the alteration may develop are yet to be explored thoroughly. One hypothesis is that *the goal of empowerment may be achieved by means of enhancement particularly when concerning vulnerable individuals and communities, such as the elderly, disabled, or any other group with specific needs.*

Empowerment is not a complex notion to understand and perceive, given its common use. However, when seen as a variable, empowerment becomes challenging to measure. According to the *Cambridge Dictionary*, empowerment is 'the process of gaining freedom and power to do what you want or to control what happens to you'. The authors of the *Cambridge Business Dictionary* highlight the agent-recipient disposition of the process, defining empowerment as 'the process of giving a group of people more freedom or rights'. This latter definition opens a space for discussion regarding the disposition where a positive result of being empowered is in its essence a consequence of an action taken by another party. The use of the passive voice indicates being on the 'receiving' side of the process, lacking agency to initiate the process but 'receiving' freedoms, possibilities, and rights.

If human enhancement is seen as a framework for empowerment and improvement – of lives, possibilities, chances, and outcomes – it should still be understood as an action accompanied by a process characterised by power imbalance. This understanding is important, especially considering the promising nature of new technologies for improving the lives of people and communities, and the consequent high stakes both in relation to the benefits, existing ethical questions, and accompanying fears and anxieties.

The ways in which the goals of empowerment, emancipation, inclusion, and participation may be achieved by means of rapid technological development are being explored by different entities, including the technological companies and local governments. Matters of ethics are being discussed and aimed to be embedded in both the surrounding discourse, including legal discussions, and technological products.

Technological devices that can be categorised as human enhancement tools in the existing debates may be viewed as both opposing 'human dignity' and potentially empowering humans and communities including vulnerable and marginalised communities (Bostrom, 2005; Kirchhoffer, 2017).

The challenge in related debates lies in reasoning that can be approached from various disciplinary angles, including philosophy, law, medicine, informatics, engineering studies, and so on. Finding the most inclusive and convincing framework for deliberation on human enhancement phenomena and tools is not an easy task. The discussion of potential benefits versus risks at this point may be characterised as even more intense than in application to other AI and broader emerging technology tools, as in the case of human enhancement it touches upon the very fundamental, ontological questions of being.

Being part of a broader debate on new technologies, their benefits and risks implies inclusivity and plurality within the debate itself, accommodating many different voices. In terms of potential benefits, this means including members of vulnerable communities who can primarily benefit from the practical technological solutions under the human enhancement umbrella. The 'ideological' polarisation of the debate can be seen as an obstacle to a truly inclusive discussion. Furthermore, the general philosophical discussion at the core of the problem may be viewed as an unnecessary complication, especially in the practical realisation – *the creation of technological tools that can enhance the physical, cognitive, and artistic abilities of people, particularly those in vulnerable positions*.

While there may be motivation, including economic incentives, for technological providers to work on solutions that, if successful, could free up resources in the healthcare field, and for governments to explore the related opportunities, individual fears and collective anxieties directed towards the technological progress may complicate these processes.

(Human) Identity and Creativity

Contemporary interdisciplinary scholarship views the process of identity-making as open-ended, with identity itself seen not as something constant and fixed, but constantly changing, with a possibility to be 'photographed' at a particular moment of time. Identities with their fluidity and metaphorical 'liquidity' – referring to Bauman's (2012) concept of 'liquid modernity' – are not easy to approach from a methodological standpoint. Discourse analysis is considered the most feasible research method for exploring these dynamic concepts (Ainsworth & Hardy, 2004; Benwell & Stokoe, 2006; De Cillia et al., 1999).

Identity is perceived on two levels: individual (personal) identity and collective (group) identity. Taylor (1989a, pp. 32–33) approaches identity through the categories of recognition and misrecognition, alienation, and othering: 'We define our identity always in dialogue with, sometimes in struggle against, the things our significant others want to see in us'. Personal identity can be influenced in both positive and negative ways. Using practical examples, misrecognition, alienation, and othering may create conditions for resentment, while recognition, on the contrary, may lead to flourishing. Of course, there are no common rules when it comes to the complexity of an individual: Taylor (1989b) explores factors that, in his view, historically had an impact on contemporary identity-making, some of them with contradictory effect. The dialectic of identity is embedded in how the matter is defined. Both recognition and misrecognition, defining individual identity, are, by their nature, contradictory vectors, predetermining the dialectic of identity on an individual level.

Creativity may be viewed as an exercise of one's personal identity. How do we better make sense of oneself, if not by means of art created by an individual? As previously outlined, the discussion on human enhancement tools and practices, potential benefits and risks, may be characterised as even more radical than the discussions on other artefacts of emerging technologies, since the matters in consideration relate to the very foundational ontological questions of human and being. The very same logic applies to the notion of creativity. Throughout its continuous history, creativity has been viewed as a distinct and exclusively human feature, with both the abilities to create art and to appreciate it regarded as 'high' characteristics of human nature. The possibility of refining the definition and the approach to creativity touches upon the fundamental questions, such as what it means to be human, hence going beyond a 'narrow' discussion on the subject.

This view brings a related question, such as whether creativity should be exclusively viewed as a characteristic of an individual – and in this way, a descriptive and probably immanent feature – or, as it is the case with identity, a continuous process that may be captured, 'photographed' at a moment in time but then changing. Alternatively, creativity can be seen as a characteristic of non-human agents, including technologies, machines, and AI systems. The answer to this question will have very practical implications, with one of the straightforward examples being legal implications, including the reasoning behind the copyright laws. How should one approach this question?

One possible approach is concept analysis aiming to determine the changes concepts have undergone as a consequence of contextual developments, including rapid technological development. Creativity is not a standalone concept that can be approached in such a way. For instance, the concept of vulnerability is now applied to non-human entities including AI systems and machine learning. For instance, as it is stressed in the *Hacking AI* report (Lohn, 2020),

> Artificial intelligence is vulnerable to cyber attacks. Machine learning systems – the core of modern AI – are rife with vulnerabilities. Attack code to exploit these vulnerabilities has already proliferated widely while defensive techniques are limited and struggling to keep up. Machine learning vulnerabilities permit hackers to manipulate the machine learning systems' integrity (causing them to make mistakes), confidentiality (causing them to leak information), and availability (causing them to cease functioning).

This is an interesting example of an important 'humane' notion being broadened to other fields and non-human subjects. This process brings the need for philosophical reflection and analysis on what exactly happened, what this process indicates, and on which existential and ontological questions we may now answer differently, keeping in mind this new evidence.

Not only is the question of whether creativity can be an attribute of machines and artificial systems relevant to the context of rapid technological development, but also the question of whether the concept of creativity itself can be reshaped by the changed context due to technological developments. It is possible that this debate may reach a middle ground by outlining a distinct notion of artificial, algorithmic, or machine creativity as specific and attributed to technological artefacts rather than to humans. In this line, 'artificial creativity' has a chance to become a concept encompassing a meaningful contrast with 'natural', human, and non-human animal creativity. The 'secondary' derivative character of AI creativity may become encoded in this term, with 'artificial', human-made, indicating both the presence of AI technologies and the distinctive, 'non-human' nature of the phenomenon.

The 'External' Machine: Societal Anxieties and Existential Risks

As discussed earlier, the human enhancement paradigm from the very beginning was focused on intervention, primarily targeting a human body with its physical, cognitive, and sensory abilities, depending on the focus and goal. The descriptive aspect of the human enhancement discourse relies on scientific imaginaries, products of creativity, be it 'individual' creativity of the author or 'collective' creativity of readers or viewers. The inward vector of human enhancement, directing improvement from the 'outside' to the 'within', was perceived as the only possibility. The design of the enhancement action could be seen as making the external internalised, becoming a feature of the human body. The 'machine' part thus played a dual, internal–external role: an outside-the-body feature incorporated into the body. This external nature made it reasonable to associate the feature itself and the entire condition of enhancement with artificiality opposing the 'natural' condition. The duality of human and artificial, with the primary, leading role of the human condition and the supporting role of the artificial condition, is inscribed into the notion itself: enhancement referring to 'human', with the 'human' being central and the enhancement condition itself being supplementary. The question then arises: Why were human enhancement features or tools not considered as external, meaning that such a feature or tool could stay 'outside' of the body, not being incorporated into the body, while serving the purpose of enhancing human abilities?

Such an 'external machine' disposition may serve as the framework for viewing AI as an enabler of human enhancement, an (artificial/non-human/machine) facilitator of this process.

What is the external machine, then? I would suggest the following definitions: *an artefact not being part of a human body but participating in the physical, cognitive, sensory, or other types of enhancement*; *an enhancement agent located separately from the enhanced individual; an artefact or a tool that has augmentation potential without the need to be internalised as a physical entity*.

The traditional focus on bodily interventions may be shifted by considering the integration of external devices, technologies, or tools that extend and enhance various aspects of human capability without the need for bodily interventions. For example, augmented reality and virtual reality technologies can be seen as external machines creating immersive environments that augment human perception of reality. These technologies can extend human experiences beyond the physical constraints of the immediate environment for purposes of entertainment, education, or training, all without the need for bodily interventions.

Future Research Agenda

The discussion on AI creativity in the context of human enhancement implies the need to address expressions of multiple fears, both on individual and societal levels, caused by the unknown. These fears do not fully correspond with attempts to rationalise the discourse, including the outlines of potential risks and ways to mitigate them; these fears also have their roots in cultural artefacts, science fiction books and movies, and other works of human imagination.

The process of personal identity shaping may involve different types of discourse. One is recognising ourselves as unique actors of creativity, the actors who nevertheless can be supported and enhanced by machines, including AI, in creative activities. Another is assessing AI creativity as a distinct, 'artificial' form of creativity different from the human and non-human animal creativity. The third is viewing ourselves as not entirely unique, sharing the very same creative abilities with machines.

The latter scenario relates to misrecognition, which is one of the important processes in the open-ended formation of identity. In this way, the individual fears and societal anxieties could be seen as predictable, understandable, and having their rationale, yet, having also the potential to disrupt – by the linguistic analogy with the disruptive character assigned to some of new emerging technologies – the processes. Multiple questions need to be further discussed to frame the discourse on possibilities and reasonable doubts, going beyond the fears and societal anxieties. In view of the points discussed above, I suggest the following questions:

RQ1: How can we facilitate a meaningful AI creativity discussion that involves experts from different fields, avoiding the dispersion of knowledge and encouraging cohesion?

The interdisciplinary approach is necessary for a meaningful discussion on AI creativity, yet it can become very formal if the 'translation' efforts are not made. The situation when experts from different disciplines interested in AI creativity simply gather for a discussion may not be sufficient. A truly rewarding and enriching interdisciplinary discussion may require thoughtful design and facilitation, where professionals are not only willing but are prepared to collaborate. Good practices, including the ways in which translation from discipline to discipline is better conducted and the shared vocabulary deliberated, need to be discussed.

RQ2: How can we overcome the ideological polarisation in the human enhancement debate and promote an interdisciplinary rationale instead?

In its historical development, the human enhancement debate tended to be bipolarised, with transhumanists and bioconservatives forming the discursive extremes. The paradox here is that while the philosophical approach has explanatory potential for the multiple questions raised by new emerging technologies, in the case of the human enhancement debate, the philosophical and idea-driven outlook is persistent and almost exclusive. This leads to the diversion of the debate from the practical path of real-life cases that need attention and elaboration to theoretical extremes. While abstraction, whether based on reasoning or imagination, is relevant to the subject matter, other, more practical, 'hands-on' and interdisciplinary approaches are also needed, as they may help answer important questions, with examples situated in the engineering, medical, legal, and other domains of knowledge.

RQ3: Can co-design of AI creative tools constitute a good enough response to fears caused by the discourse on AI creativity and AI creativity itself?

The rapid development of technologies causes both opportunities and fears. One direction of research is the future of work, including creative fields. It is important to determine the best ways in which the discussions can be facilitated and anxieties tackled. Collaborative thinking and reflection may ease the minds of creative professionals and bring forth interesting solutions. It would be useful to explore the opportunities of co-creation: bringing together technological providers and creative specialists may lead to the development of inclusive solutions and at the very same time strengthen the human enhancement framework of AI tools.

Funding

This chapter has been supported by the EU's Horizon Europe research and innovation programme under grant agreement No 101070325 (Innovative Solutions Responding to the Needs of Cities & Communities – CommuniCity).

References

Ainsworth, S., & Hardy, C. (2004). Critical discourse analysis and identity: Why bother? *Critical Discourse Studies*, *1*(2), 225–259.

Almeida, M., & Diogo, R. (2019). Human enhancement: Genetic engineering and evolution. *Evolution, Medicine, and Public Health*, *2019*(1), 183–189.

Bauman, Z. (2012). *Liquid modernity*. Polity Press.

Bavelier, D., Savulescu, J., Fried, L. P., Friedmann, T., Lathan, C. E., Schürle, S., & Beard, J. R. (2019). Rethinking human enhancement as collective welfarism. *Nature Human Behaviour*, *3*(3), 204–206.

Benwell, B., & Stokoe, E. (2006). *Discourse and identity*. Edinburgh University Press.

Bostrom, N. (2005). In defense of posthuman dignity. *Bioethics*, *19*(3), 202–214.

Colton, S., Mántaras, R. L., & Stock, O. (2009). Computational creativity: Coming of age. *AI Magazine*, *30*, 11–14.

Cornips, L., & van Asselt, M. (2013). Human enhancement in futures explorations. In B. Koops, C. Lüthy, A. Nelis, C. Sieburgh, J. Jansen, & M. Schmid (Eds.), *Engineering the human* (pp. 45–61). Springer.

De Cillia, R., Reisigl, M., & Wodak, R. (1999). The discursive construction of national identities. *Discourse & Society*, *10*(2), 149–173.

Eapen, T. T., Finkenstadt, D. J., Folk, J., & Venkataswamy, L. (2023). How generative AI can augment human creativity: Use it to promote divergent thinking. *Harvard Business Review*. Retrieved December 3, 2023, from https://hbr.org/2023/07/how-generative-ai-can-augment-human-creativity

Juengst, E. T. (1997). Can enhancement be distinguished from prevention in genetic medicine? *The Journal of Medicine and Philosophy*, *22*, 125–142.

Kaufman, A. B., & O'Hearn, W. J. (2017). Creativity in non-human animals. In J. C. Kaufman, V. P. Glăveanu, & J. Baer (Eds.), *The Cambridge handbook of creativity across domains* (pp. 492–506). Cambridge University Press.

Kirchhoffer, D. G. (2017). Human dignity and human enhancement: A multidimensional approach. *Bioethics*, *31*(5), 375–383.

Lohn, A. (2020). *Hacking AI: A primer for policymakers on machine learning cybersecurity*. Center for Security and Emerging Technology. Retrieved December 3, 2023, from https://cset.georgetown.edu/publication/hacking-ai/

Lüthy, C., & Koops, B. J. (2013). Towards homo manufactus? An introduction to this volume. In B. Koops, C. Lüthy, A. Nelis, C. Sieburgh, J. Jansen, & M. Schmid (Eds.), *Engineering the human* (pp. 1–9). Springer.

Nelis, A., & Posthuma, D. (2013). Genetic enhancement of human beings: Reality or fiction? In B. Koops, C. Lüthy, A. Nelis, C. Sieburgh, J. Jansen, & M. Schmid (Eds.), *Engineering the human* (pp. 63–70). Springer.

Nikkel, B. (2023). Human vs. machine: A comparative study on creativity in the age of AI. *Deepgram*. Retrieved December 3, 2023, from https://deepgram.com/learn/ai-beat-human-creativity

Svana, L. (2017). (Military) human enhancement – ethical aspects. *Human Affairs*, 27(2), 155–165.

Taylor, C. (1989a). The politics of recognition. In A. Gutmann (Ed.), *Multiculturalism: Examining the politics of recognition* (pp. 25–73). Princeton University Press. Retrieved December 3, 2023, from http://fs2.american.edu/dfagel/www/Class%20Readings/Taylor/multiculturalism.pdf

Taylor, C. (1989b). *Sources of the self: The making of the modern identity*. Harvard University Press.

Part 3

Industry-Specific Discussions

Chapter 12

Make Some Noise

Digital Art and Artificial Intelligence Striking a Chord

Trevor Borg, Elena Said, Jonathan Attard, and Dylan Seychell

Introduction

> *The world of perception, or in other words the world which is revealed to us by our senses and in everyday life, seems at first sight to be the one we know best of all. For we need neither to measure nor to calculate in order to gain access to this world and it would seem that we can fathom it simply by opening our eyes and getting on with our lives. Yet this is a delusion.*
>
> (Merleau-Ponty, 2002, p. 31)

Merleau-Ponty's words appropriately capture the scope of the current digital art and artificial intelligence (AI) collaborative project titled 'Make Some Noise' – a project that connects human perception with machine programming. The project discussed here brings the subjectivity of human senses and scientific objectivity together, as one feeds and reconfigures the other with the aim of establishing a liminal space that belongs to both of them and to none. Departing from a combined aesthetic and eco-critical standpoint, 'Make Some Noise' aims to translate an environmental concern into an artefact with the help of AI algorithms. Before delving deeper into the project, we have to first investigate the context that provoked this critical project, which is central to the ensuing environmental commentary.

The project originates in Malta, a tiny but historically loaded Mediterranean island equidistantly positioned between two continents, Europe and Africa. Due to its strategic location Malta has always played an important role in the course of history. Some of the world's earliest free-standing structures, dating back 5,000 years, are located on Malta and Gozo, a sister island situated a few kilometres away from Malta. Spanning only 316 km^2, Malta's sun-kissed landscape is embellished with history, beautiful beaches, and pristine cliffs, but it is also heavily weighed down by limestone and concrete buildings. Tower cranes are ubiquitous wherever you go in Malta and Gozo. Echoing Merleau-Ponty's words, this reality is revealed to us through the senses; we can perceive it by merely turning our head around and approaching the world through our bodies.

DOI: 10.4324/9781003453901-16

The authors of this chapter acknowledge the need to resort to 'less' anthropomorphic measures, which may include human–computer interactions, to further highlight some of the urgent specificities in our changing environment (Grosz, 2008, p. 6). These are urgencies that necessitate stories (Haraway, 2016, p. 37); they must be permitted to sound-off, to be heard and listened to. This situation precipitated a creative AI project fashioned on environmental 'activism' that is meant to generate further awareness through poetic-political intertwining. The project, which developed from a Master's in Fine Arts in digital arts research at the University of Malta titled 'DIN l-Art Ħelwa' (2023), echoes a collective perception; it resonates with the current state of affairs, while it also opens up a 'world of its own' (Merleau-Ponty, 2002, p. 71). Often, it is necessary to introduce further *noise* to make one's self heard, allowing chaos more autonomy so as to effectively rise above the deafening 'din' (Grosz, 2008, p. 7).

As a group of artists and computer scientists, we are particularly interested in advancing our ecological preoccupations through digital art and AI. In this case 'machines' have been used to investigate 'other' 'machines'. As Donna Haraway puts it, '[i]*t matters what stories tell stories*' (2016, p. 35); in this event the aim is to establish a method of telling a story, co-authored with the help of algorithms, that imbricates perception with facts. Timothy Morton (2009, p. 8) argues that art's ability to question reality could be very telling about the state of our environment; art deals with 'being and seeming', it incites us to think differently (p. 10). 'Make Some Noise' is designed to investigate how the 'what is' might become when reterritorialised into a totally different context, reliant on object detection algorithms. Although the AI translation is based on supervised learning methods, designed to be objective and which rely entirely on the training data they receive, there is still a strong element of novelty involved. Even if it is common knowledge that '[t]he ability to allow some questions to remain unanswered' and to be able 'to tolerate ambiguity' are among the characteristics that distinguish humans from machines (King, 2015, p. 31), there might still be an element of 'subjectivity' which could lead to very interesting results. This chapter discusses how a digital art and AI amalgam might generate unpredictable results through deterministic programs with set parameters, including confidence threshold and non-max suppression.

The co-creative project is divided into two parts which can be described as '*Collecting*' and '*Composing*', respectively, whereby the first part feeds directly into the second. Both parts are stringed together by algorithms that objectively compute confidence levels based on learned patterns, which can be tweaked to produce 'random' results and novel sound iterations of the Maltese national anthem. Echoing Alan Turing's ideas of transforming the machine into an unpredictable 'thinking' entity, the project considers subjectivity as a provisional aspect based on 'durational time' (King, 2015, p. 46).

Collecting

'Make Some Noise' emerges from the simple act of walking in six selected villages in Malta and Gozo. Walking fosters a profound connection with the environment, allowing for an immersive engagement with the place and the landscape, as emphasised by Solnit (2002). The first part of the project involves walking and documenting tower cranes dotting the Maltese landscape in six different villages. The decision to centre the study on the number 'six' was motivated by the number of verses that make up *Lil Din L-Art Ħelwa*, the Maltese national anthem written in the form of a prayer. Authored by a priest (Dun) Karm Psaila, and composed by Robert Samut, *Lil Din L-Art Ħelwa* became the national anthem of Malta following the islands' independence from the British in 1964, though the lyrics and music originated in 1922. Aligned with the previously discussed political dimension, the choice of a national anthem underscores the project's core environmental message. The lyrics of the Maltese anthem are reproduced below:

Lil din l-art ħelwa, l-Omm li tatna isimha,
Ħares, Mulej, kif dejjem Int ħarist:
Ftakar li lilha bil-oħla dawl libbist.
Agħti, kbir Alla, id-deħen lil min jaħkimha,
Rodd il-ħniena lis-sid, saħħa 'l-ħaddiem:
Seddaq il-għaqda fil-Maltin u s-sliem.
(Dun Karm Psaila, 1922)

O Little Island, Lord, as ever Thou hast guarded!
This Motherland so dear whose name we bear!
Keep her in mind, whom Thou hast made so fair!
May he who rules, for wisdom be regarded!
In master mercy, strength in man increase!
Confirm us all, in unity and peace!
(English Translation by May Butcher)

In an effort to gain a comprehensive understanding of the evolving landscape the choice of localities spans Malta and Gozo from North to South as follows: Malta – Rabat, St Julians, Msida, Żurrieq; and Gozo – Qala, Sannat (Figure 12.1). The designated locations correspond with the composition of the anthem. Six viewpoints were eventually established in each location to capture the same section of the landscape upon each visit. Every location was revisited once a month for six consecutive months, totalling a final selection of 36 photographs – six photographs of each separate location. The aim behind the repeat visits is to comprehensively capture the changing landscape over a period of time so as to present the algorithm with more data to work with.

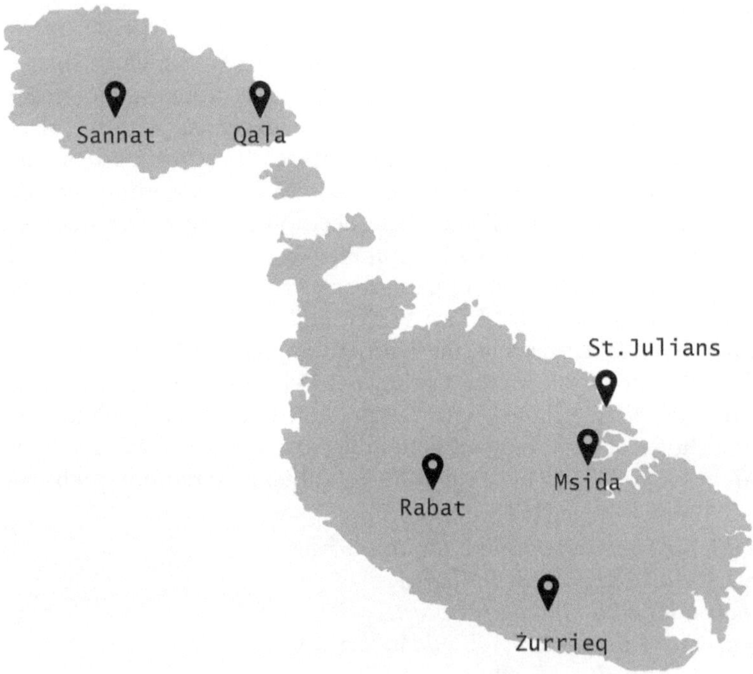

Figure 12.1 The six locations from which the images were taken as shown on Google Maps.

Research in local newspapers and through phenomenological enquiry informed the identification of sites in Malta and Gozo, undergoing substantial change and development (Borg, 2023). As James Gibson asserts (1979/2015, pp. 187–188):

> *Seeing the world at a travel[l]ing point of observation, over a long enough time for a sufficiently extended set of paths, begins to be perceiving the world at all points of observation, as if one could be everywhere at once.*

Walks took place in specific locations strategically chosen to achieve across-the-board coverage of building sites spanning Malta and Gozo. Despite an initial excess of options, site visits were conducted prior to starting the documentation process to evaluate different viewpoints, resulting in a selection of locations offering unobstructed views of tower cranes in the landscape. As will be discussed in the next section, it is important to frame the landscape and the objects appropriately, in order to render the algorithm's identification method more precise and effective. This process of intentional navigation resonates with principles of psychogeography as practised by the Situationists International (Sadler, 1998/1999).

Multiple images were captured with a mobile phone camera to allow for subsequent selection of images pertaining to each location. Prior to the actual image collection process, an effort to establish a trace of breadcrumbs was made by activating geolocation to ensure precise documentation of tower cranes in the landscape (Figure 12.2). Following the image collection procedure, for comparative purposes the categorisation process has been based on the location and the time of year when each image was taken. The six photographs of the individual locations were selected on the basis of clarity and tower crane visibility.

By encountering the 'audible' representation of cranes in the Maltese landscape, the study aims to amplify their impact on a different sensorial level. This rationale, influenced by John Cage's philosophy (1961), emphasises the imperative of sound creation being aligned with its essential purpose. The cacophony of visual noise will be used to reconfigure the anthem in ways that resound with the changing landscape. Cage's ideas in relation to 'silence' (1961) resonate with instances where the AI's 'confidence' in identifying an image as a tower crane is low, or when it struggles to recognise cranes in the images. Cage's suggestion that there is no 'true' silence aligns with the algorithm's unsuccessful identification of a crane when a crane is present in the image. In this case, 'silence' is understood as the absence of any alterations in the original composition of the national anthem. 'Silence' may serve as an added interpretation of ambiguous visual data so as to introduce further depth and complexity to the auditory experience. The placing of 'silence', or white noise, as part of the general altered composition could be used in lieu of incomplete object identification instances. This is also comparable to Derrida's (1976) notion of the 'fissure', as he argues that intervals, stoppages, and gaps are essential to song and similarly to music.

Drawing again on Derrida (1976), the process of altering a national anthem can be considered as a movement of deconstruction, whereby the aim is not to dismantle the outward appearance of an existing composition but to change it from within. In *Ecology without Nature* (2007), Morton draws inspiration from the trajectory of deconstruction, which seeks to uncover contradictory elements within structures of meaning (p. 6). This stems from Derrida's deconstructionist theory, which involves breaking down elements to grasp their true essence. Borrowing from Derrida's notion of dismantling hierarchical thought systems (Norris, 1987), this entails a re-examination of the visual content, focusing on the 'noise' represented by the tower cranes embedded in the landscape. The image emerges as a form of text that eventually forms a bond with the sound of the national anthem (Derrida, 1976).

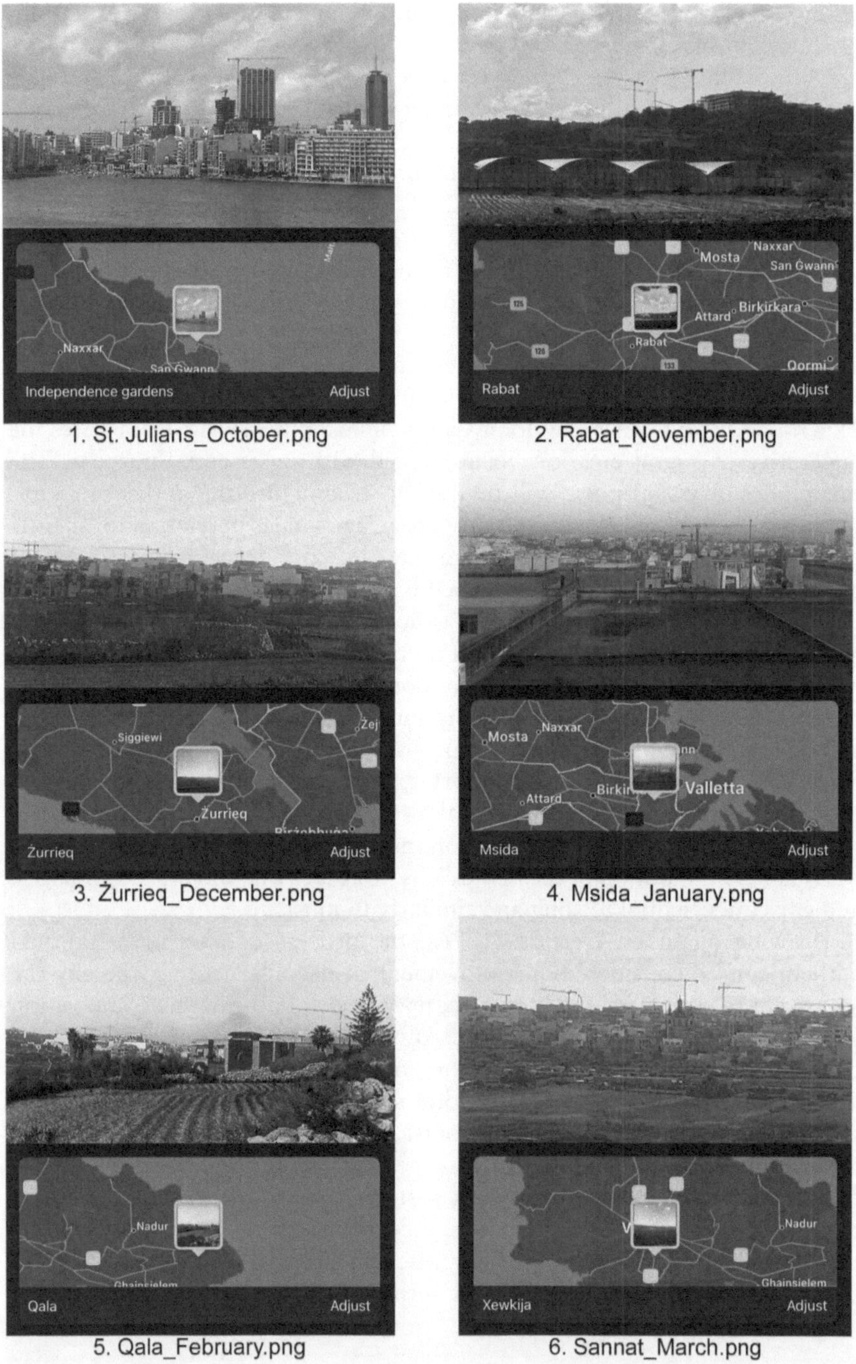

Figure 12.2 A photograph of each locality with its geolocation shown on the Maltese map.

The In-Between Stage: Detecting

Exploring specific locations through walks allowed for the identification of emerging landscape patterns. The 'changing' observations were recorded photographically, clearly visible to the naked eye, yet difficult for a human to track and monitor across different images. While this can prove challenging for humans, it is an achievable task for machines trained to detect such changes. Machine learning is the largest branch of AI and includes programs which can model a representation of the world. Over time, with more examples, these models improve and provide a more detailed interpretation of the environment in which they are trained (Ekman, 2022). Specifically, in this project, supervised learning algorithms were used, showing 807 images (579 for training; 150 for validation; 78 for testing) featuring labelled examples of cranes. These were annotated with bounding boxes to identify where in the image the algorithm could find a crane. By analysing the pixel patterns, colours, textures, and other features distinguishable to the computer, the algorithm can learn to recognise when a crane is present in the image.

The process of harnessing location-specific images to manipulate audio consisted of two pivotal phases. Initially, AI was utilised to extract pertinent data from images linked to specific localities. Subsequently, extracted information becomes instrumental in introducing distortions to the corresponding audio content, which will be explored in the next stage. Figure 12.3 presents a data flow diagram of the process.

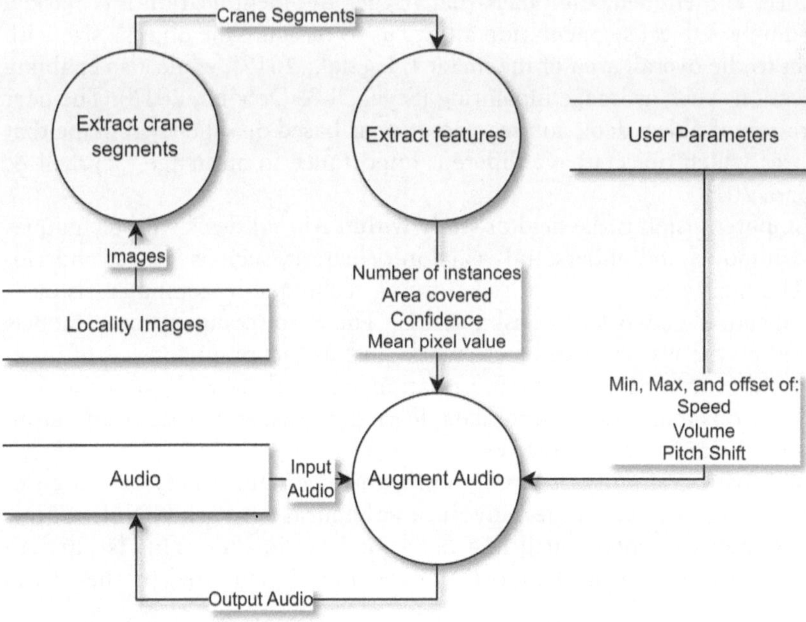

Figure 12.3 Data flow diagram illustrating the how location-based images are used to distort audio.

The approach commenced with the extraction of location-specific features from the images. To accomplish this, the YOLOv5-Segmentation model (You Only Look Once) was employed, renowned for its prowess in object segmentation. The selected features from each image were then averaged for each locality, encompassing; 'Number of cranes', 'Crane area coverage ratio', 'YOLO model confidence score', and 'Mean pixel value of the image'. The extraction process involves feeding each image into the YOLO model to predict crane segments. Consequently, the number of cranes per image was tallied followed by the calculation of the crane area coverage ratio using a generated mask, the calculation of the average YOLO model confidence scores for each crane prediction, and the computation of the mean pixel value for the entire image. These location-specific characteristics, distilled from all images in each locality, constitute the foundation for audio distortion (Figure 12.4).

The model compares new unlabelled images to its trained representation and predicts whether unseen images contain cranes. In machine learning, this is known as *classification*, where the predominant class in the image, according to the training of the model, is predicted. The next level is object *detection*, where the model not only recognises the main class in the image but also draws a bounding box to show which area of pixels represents the class it is perceiving. This implies that the bounding box contains pixels that represent the object and other pixels that happen to be within the bounding box due to its rectangular nature. Object detection is followed by object segmentation, which identifies objects at a pixel level. This means that when an object is identified, the pixels that trigger the identification are marked accordingly. Object segmentation allows us to measure the object's size with respect to the overall area of the image (He et al., 2017), while also enabling applications such as image inpainting (Seychell & Debono, 2016). The next generation of object detection segmentation is based on saliency ranking that indicates which objects have different importance in an image (Seychell & Debono, 2018).

Computer vision is the field of study within AI that deals with techniques that can model and understand visual information, such as images and videos. The most popular supervised learning technique in computer vision is that of neural networks (Szeliski, 2022). These are computational models inspired by the way human brains work. Just as our brain consists of interconnected neurons that process information, a neural network comprises layers of neurons that process input data, learn patterns, and produce an output based on the training they receive.

The most commonly used configuration of a neural network in computer vision is known as the convolutional neural network (CNN). CNNs are designed to process grid-like data, such as images. This is particularly useful in detecting the mostly linear 'noise' generated by the cranes

Figure 12.4 High-level graphical representation illustrating the utilisation of Yolov5-Segmentation and simple averaging for extracting location-specific features.

projected over the Maltese skyline. In the context of this project, this also means that the algorithm can recognise the curve of a crane's hook or the straight lines of its mast. By recognising these patterns and hierarchies of features, CNNs can identify objects within an image with accuracy that is proportional to the number of training examples it would have seen (Lakshmanan et al., 2021). Going back to the 'subjective' element that delineates art from science, there are still 'grey areas' which the algorithm might not be able to easily identify or outright dismiss as 'non-cranes'. The YOLO object detection technique (Redmon & Farhadi, 2018) is a modern object detection and segmentation model that uses CNNs. YOLO, while building on the principles of neural networks, offers a unique approach to object detection, making it particularly suited for our landscape analysis.

YOLO's unique approach to object detection mirrors, in some ways, our own human perception. When we glance at a scene, we do not analyse it in segments or parts; we take in the entire view at once, identifying objects and patterns (Krieman, 2021). Similarly, YOLO analyses an entire image in a single pass, making it both fast and efficient.

The mechanics of YOLO involve dividing the image into a grid (Lakshmanan et al., 2021). Each cell of this grid predicts if it contains the centre of a bounding box and the dimensions of this box (Redmon & Farhadi, 2018). It is akin to how our eyes might focus on specific points of interest in a scene, even though we are taking in the whole view or instance of the image. The model then assigns a class label, such as 'crane', to each bounding box, along with a confidence score indicating the probability of the prediction being correct. This level of confidence is also related to the training examples seen by the network. If input includes a crane that is similar to several cranes seen during training, then the level of confidence would be high because the network can recognise it well. On the other hand, the confidence level is low if the crane detected in the input image does not resemble the cranes in the training examples. This implies that there might be another object that looks like a crane but is not a crane that is incorrectly identified as one. In this case, such ambiguity is a desirable factor since it introduces an element of 'subjectivity' that bridges the gap between art and AI. Just as our brain filters out unnecessary information to focus on what is essential, YOLO disregards bounding boxes with low confidence scores (Ekman, 2022). This ensures that only the most probable predictions are retained. The result is a model that can efficiently and accurately identify objects like cranes in input images, even if they are partially occluded or overlap with other objects.

The integration of the YOLO object detector in our project brings together human perception and machine precision. By mimicking certain aspects of human perception, YOLO provides a robust and efficient solution to our landscape classification experiment.

Composing

Art '*gives voice to what is unspeakable elsewhere*' (Morton, 2009, p. 12). Through this project, 'voice' is being translated into a chaotic symphony aimed to attract further attention to the plight of the environment. The second part of the process, referred to as 'Composing', involves the application of numeric data based on crane identification to the sound composition of the Maltese national anthem. Transformations applied to the anthem are again determined by an algorithm, a different one from that used for identifying the objects discussed in the previous section.

Three dynamic elements were introduced to the original audio, namely: Volume, Speed, and Pitch, inspired by Russo (Russo, 2023). These modifications were influenced by tower crane image data, leading to increased intricacies as more images were incorporated, with every image introducing its own audio segment change. Each location-based feature assumes the role of a modifier within the audio. This transformative process unfolds through several steps:

1. **Audio Segmentation:** The audio was divided into clips, with each clip representing a specific locality. These clips were weighted based on the number of cranes present in each locality, meaning that localities with more cranes had a longer clip duration. The order of the clips depended solely on the input sequence of the localities.
2. **Distortion Parameters:** For each locality clip, the average area coverage, the average segment confidence, and the mean pixel value were used as parameters to distort the audio.
3. **Clip Reassembly:** After applying distortion to each clip, clips were reassembled to complete the audio distortion process.

To allow for more flexibility, specific boundaries may be defined, including minimum, maximum and offsets for each modifier. These boundaries facilitate the use of location-based features as a percentage within predefined ranges. The introduction of offsets caters for scenarios where calculated values venture outside acceptable limits. For instance, exceptionally low volume levels could render the audio nearly inaudible. The crane area coverage ratio influences speed, the YOLO model confidence score affects volume, and the mean pixel value of the image influences the pitch of the clip. Equations were

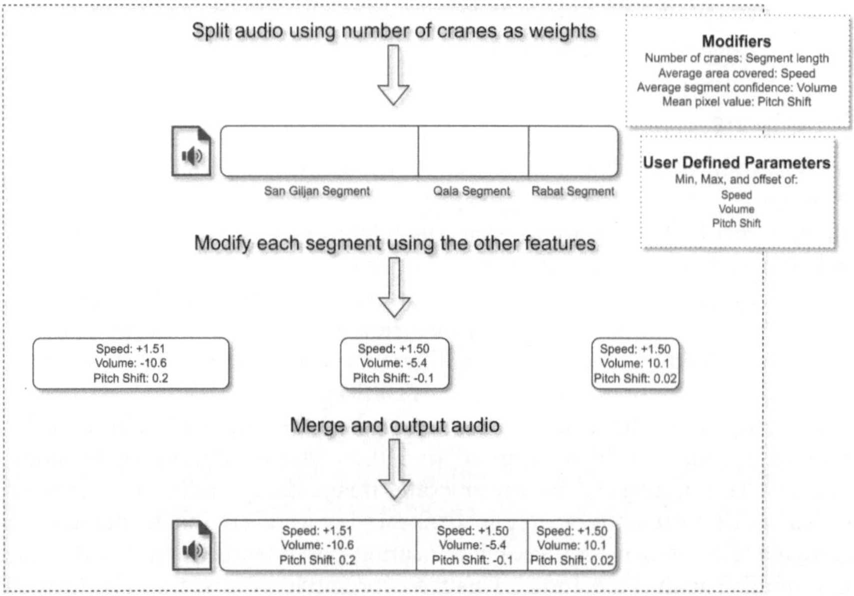

Figure 12.5 High-level graphical example showing how location-based features distort audio.

employed for each modifier as shown. A graphical representation of this process can be seen in Figure 12.5.

$$\text{New\_Speed} = \text{Previous\_Speed} \times (\text{Minimum} + \text{Average (Area ratio)} \times (\text{Maximum} - \text{Minimum}) + \text{Offset})$$

$$\text{New\_Volume} = \text{Previous\_Volume} + (\text{Minimum} + \text{Average (Confidence ratio)} \times (\text{Maximum} - \text{Minimum}) + \text{Offset})$$

$$\text{New\_Sample\_Rate} = \text{Previous\_Sample\_Rate} \times 2 \wedge \text{Minimum} + \text{Average (Mean pixel / 255)} \times (\text{Maximum} - \text{Minimum}) + \text{Offset})$$

Equation for altering pitch using sample rate (Yildirim, 2021).

The Mel-Spectrograms in Figure 12.6 provide a detailed visual representation of distortions in the Maltese anthem, capturing variations across locations in terms of amplitude, pitch, and frequency content. In contrast, Figure 12.7 utilises waveforms to emphasise amplitude and temporal characteristics. These two representations (McFee et al., 2015) complement

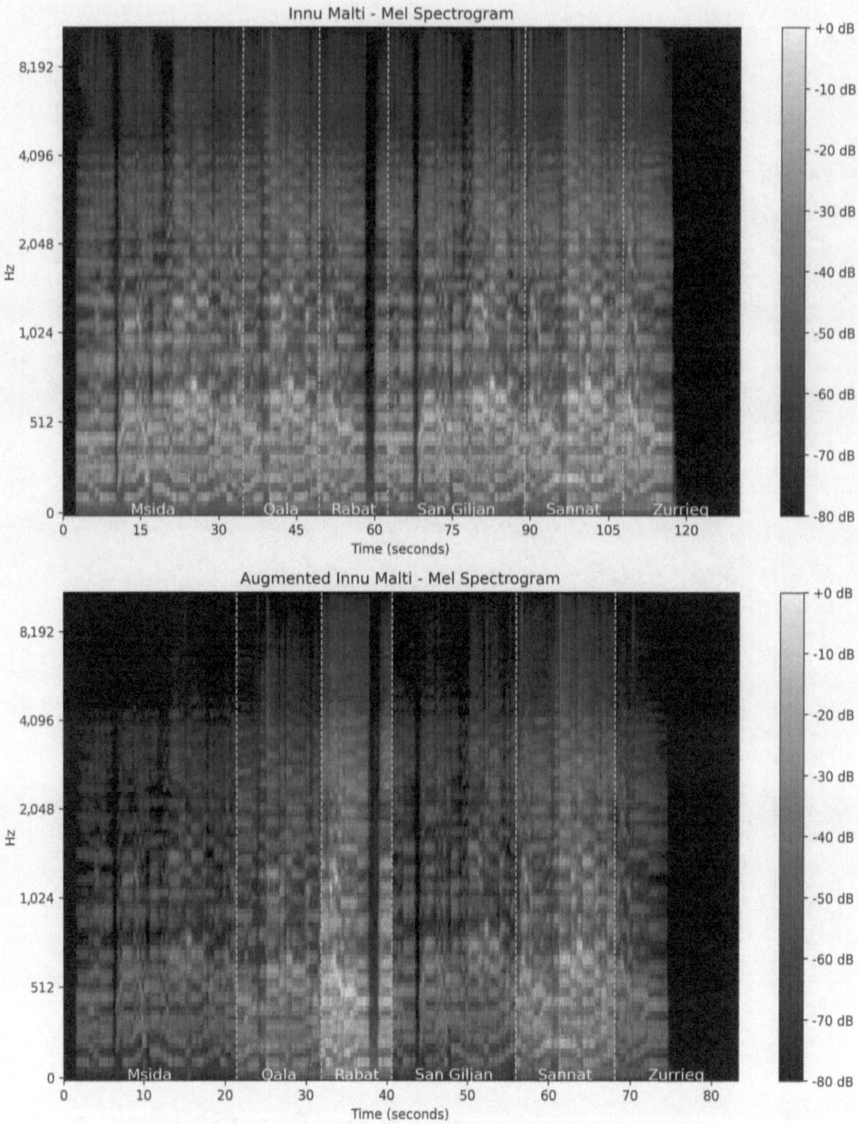

Figure 12.6 An illustrative example of distorted Maltese anthem using location-specific images. The Mel-Spectrograms depict the contrast between the original audio and the augmented version. In these Mel-Spectrograms, each section is partitioned into location-based clips, providing a visual insight into the variations between each locality and the original audio.

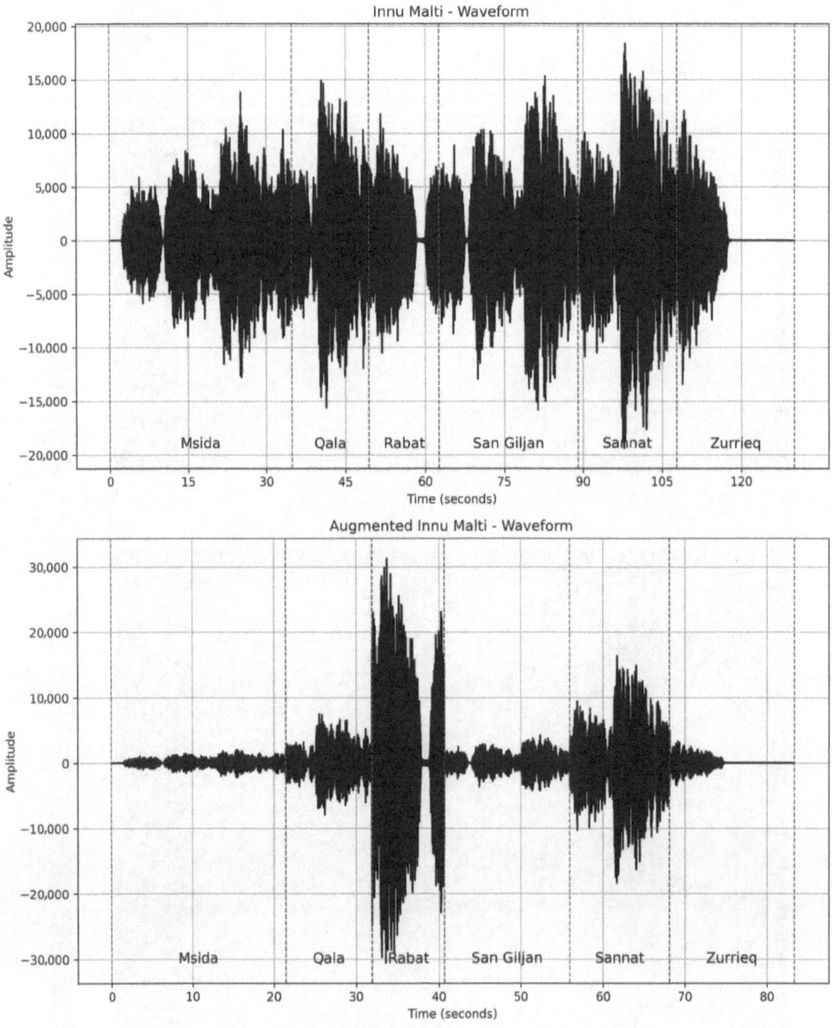

Figure 12.7 Comparative waveform analysis of distorted Maltese anthem. This figure presents waveform representations, akin to those in Figure 12.6, to illustrate the disparities between the original audio and its augmented version. Each section is subdivided into location-specific clips, providing a visual depiction of the differences across various localities compared to the original audio.

each other by offering insights from both spectral and temporal viewpoints, enhancing our understanding of audio alterations across different localities. The hybrid digital art and AI project opens up a space for diverse creative opportunities and intermedia collaborations. Throughout the process and development of 'Make Some Noise', we became aware of the uncharted territory that is still waiting to be discovered and explored in the near future.

Further Research Agenda

The current chapter describes an experimental amalgam between digital art and AI within an eco-critical context. Two separate algorithms have been used to deal with visual and sound data. There are multiple approaches which can be taken in order to further complicate the objectives of this co-creative project and to allow art and AI to engage more fully in conversation with each other. Here we are listing three possible research questions that can be taken up to elevate the project to a more sophisticated level:

RQ1: How could an automated selection of places arise from newspaper articles reporting on construction and the environment?

In 'Make Some Noise', newspapers have been manually reviewed; a process which is time-consuming and which needs to be repeated at specific intervals to include new developments emerging in the landscape. This research question considers the utilisation of AI as a creative tool to gather real-time information about the environment and to produce artworks aimed at raising environmental awareness based on information published in newspapers. This approach opens up a promising space for further creative exploration and hybrid practices. This could involve writing an algorithm with natural language processing capabilities to better understand nuances in articles, such as distinguishing between positive and negative sentiments or identifying subtle shifts in public opinion over time. Through this approach, the temporal trends arising from newspaper reports can be reviewed and assessed; by examining how construction and environmental concerns have evolved over years the algorithm may identify critical periods or pivotal events that are shaping the landscape. These contextual data could then tell us more about specific locations and strategies, taking into account long-term environmental transformations.

RQ2: How might the selection of locations be enhanced though the implementation of an algorithm capable of searching Google Earth images or photographs uploaded online by the country's official planning authorities?

Further research could also be aimed at exploring heavily constructed locations through an algorithm equipped to search for Google Earth images or photographs available online on the lands and planning authorities' websites. For this, an image search algorithm with the ability to pinpoint relevant images and extract valuable geospatial information can be explored.

Moreover, exploring the integration of these geospatial data sources with other environmental datasets could offer a more comprehensive understanding of the selected locations. The combination of Google Earth images with real-time environmental data can be investigated in relation to more precise and context-aware location choices to encourage further environmental projects. Ethical considerations are also crucial. Investigating the ethical implications of using publicly available images for location selection, including data protection and potential biases in image data, would entail a significant amount of research prior to embarking on the actual project.

RQ3: How analogous is the confidence level of AI to art's subjectivity?

The third prospect arising from 'Make Some Noise' concerns the intersection of art and AI, based on a deeper examination of the distinction, or rather indistinction, between the artist and the machine, in relation to the algorithm's confidence level in the process of object detection. The confidence level in AI detections represents the algorithm's degree of certainty or accuracy in appropriately identifying or classifying an object or data point, typically expressed as percentage. Lack of 'certainty', or the presence of 'uncertainty' in the algorithmic process, may be compared to how humans perceive the world, consequently pushing the machine closer to subjective aesthetic judgement. While our project resonates with Donna Haraway's vision of a cyborg that blurs the boundaries between human and machine in the creation of a collaborative artwork (1985/2016), further research could focus on the similarity between the two as individual agencies.

References

Borg, N. (2023, October 4). A home for every 1.75 people: Malta's construction boom in numbers. *Times of Malta*. Retrieved October 4, 2023, from https://timesofmalta.com/articles/view/27-apartments-every-house-malta-construction-boom-numbers.1058853?fbclid=IwAR1deorAFplCjlJx5XGQI7M33ofTLp1hOyMEU-ez0fpkN4t5DFnL2ERC9hg

Cage, J. (1961). *Silence*. Wesleyan University Press.

Derrida, J. (1976). *Of grammatology*. The Johns Hopkins University Press.

Ekman, M. (2022). *Learning deep learning*. NVIDIA. Addison-Wesley.

Gibson, J. J. (2015). *The ecological approach to visual perception*. Taylor & Francis. (Original work published 1979)

Grosz, E. (2008). *Chaos, territory, art*. Columbia University Press.

Haraway, D. J. (2016a). A cyborg manifesto. In D. J. Haraway & C. Wolfe (Eds.), *Manifestly Haraway* (pp. 3–90). University of Minnesota Press. (Original work published 1985)

Haraway, D. J. (2016b). *Staying with the trouble*. Duke University Press, Harvard University Press.

He, K., Gkioxari, G., Dollár, P., & Girshick, R. (2017). *Mask R-CNN 2017* (pp. 2980–2988). IEEE International Conference on Computer Vision (ICCV), Venice, Italy.

King, H. (2015). *Virtual memory*. Duke University Press.

Krieman, G. (2021). *Biological and computer vision*. Cambridge University Press.
Lakshmanan, V., Gorner, M., & Gillard, R. (2021). *Practical machine learning for computer vision*. O'Reilly.
McFee, B., Raffel, C., Liang, D., Ellis, D. P., McVicar, M., Battenberg, E., & Nieto, O. (2015, July). *librosa: Audio and music signal analysis in python*. In *Proceedings of the 14th Python in Science Conference* (p. 18). SciPy.
Merleau-Ponty, M. (2002). *The world of perception*. Routledge.
Morton, T. (2009). *Ecology without nature: Rethinking environmental aesthetics*. Harvard University Press.
Norris, C. (1987). *Derrida*. Harvard University Press.
Redmon, J., & Farhadi, A. (2018). *YOLOv3: An incremental improvement. ArXiv*. https://arxiv.org/pdf/1804.02767
Russo, M. (2023). Sonification 101: How to convert data into music with python. *Medium*. Retrieved September 17, 2023, https://medium.com/@astromattrusso/sonification-101-how-to-convert-data-into-music-with-python-71a6dd67751c
Sadler, S. (1999). *The situationist city*. MIT Press. (Original work published 1998)
Said, E. (2023). *DIN l-art Ħelwa: An ecocritical exploration of the Maltese landscape through the translation of data using machine learning methods*. University of Malta.
Seychell, D., & Debono, C. (2016). *Monoscopic inpainting approach using depth information 2016*. 18th Mediterranean Electrotechnical Conference (MELECON). https://www.um.edu.mt/library/oar/handle/123456789/16506
Seychell, D., & Debono, C. (2018). *Ranking regions of visual saliency in RGB-D content 2018*. International Conference on 3D Immersion (IC3D). https://www.researchgate.net/publication/331606565_Ranking_Regions_of_Visual_Saliency_in_RGB-D_Content
Solnit, R. (2002). *Hollow city: The siege of San Francisco and the crisis of American urbanism*. Verso.
Szeliski, R. (2022). *Computer vision: Algorithms and applications*. Springer.
Yildirim, B. (2021, February 3). How to pitch shift in Python. *Medium*. Retrieved September 7, 2023, from https://batulaiko.medium.com/how-to-pitch-shift-in-python-c59b53a84b6d

Chapter 13

Exploring Perceptions and Usage of Generative Artificial Intelligence

An Empirical Study among Management Students

Tobias Naujoks, Benjamin Schulte, and Cansu Hattula

Introduction

Artificial intelligence (AI) seeks to develop computational systems that simulate human intelligence and perform tasks such as learning and reasoning (Nilsson, 2014; Shi & Zheng, 2006). These systems can be broadly classified into two categories: analytical AI (AAI) and generative AI (GenAI).

AAI is the traditional type of AI that uses existing data to derive predictions and automations, that is AAI systems aim at recognizing patterns and making decisions based on structured data they have learned from. They need significant human effort and guidance as they follow predefined rules and algorithms. Therefore, AAI systems are generally rather transparent and explainable. They are mostly used for applications such as predictive analytics, image, and speech recognition.

GenAI, on the other hand, is an emerging form of AI that concentrates on generating creative, innovative, and realistic data or content, based on the information it has previously analysed (McKinsey, 2023). It requires less human involvement because these systems rely on unsupervised and semi-supervised learning algorithms. These algorithms learn patterns and structures in the input data without being dependent on explicit labelling or instructions. The use cases of GenAI span across various fields and domains as these systems can be used for different applications, including generating text, images, video, or code (Agrawal et al., 2022).

One famous example of GenAI is ChatGPT. The chatbot ChatGPT, which stands for Chat Generative Pretrained Transformer, was released for public use by OpenAI in November 2022 and marked a significant leap in GenAI. It represents more than just technological hype, as evidenced by the rapid adoption of the technology (Snider, 2023). With an estimated 100 million monthly active users in just two months after launch, ChatGPT is the fastest-growing consumer application in history (Hu, 2023). The adoption was, in part, facilitated by its user-friendly interface, free access, and consequently low barriers to entry, which made it accessible to a large audience.

DOI: 10.4324/9781003453901-17

In this context, the chapter aims to investigate how university students in the field of management perceive and use GenAI.

Literature Review

A comprehensive literature review was performed to determine the current advancements in AI in education (AIEd), along with a focus on the perception and usage of university students regarding GenAI.

AI in Education

AIEd has been a field of scientific research for more than 30 years, with a notable surge in global interest and publication of AIEd articles over the past decade (Chen et al., 2022; O'Shea & Self, 1986; Ouyang & Jiao, 2021; Zawacki-Richter et al., 2019). Today, AIEd represents a primary area in educational research, reshaping not only education but also knowledge, cognition, and culture (Hwang et al., 2020).

On the one hand, AI technology has left a profound imprint on teaching and learning (United Nations Education Scientific and Cultural Organization, 2019). It has permeated various facets of the education sector, promising to revolutionize education in several ways (Sharma et al., 2019; Roll & Wylie, 2016). Chen et al. (2020) delineate three general functions of AI in educational scenarios: administration, instruction, and learning. AI enhances efficiency and effectiveness by automating administrative tasks, aids teachers as teaching assistants, and delivers enriched learning experiences through customization and personalization of content tailored to individual students' needs (Chiu et al., 2023). Furthermore, AI offers step-by-step guidance, timely individualized feedback, and fosters collaborative learning and research among students (Hsu & Ching, 2023). Finally, AI helps to dismantle barriers and improve access to education and learning (Chen et al., 2020).

On the other hand, various concerns such as issues of data privacy, bias, discrimination, and ethical considerations are outlined (Hsu & Ching, 2023). Additionally, there is a growing concern about a decline in academic integrity due to the increased potential for cheating facilitated by AI (Chen et al., 2020). Some researchers emphasize that excessive reliance on AI may reduce students' interaction with peers and hinder the development of critical skills such as metacognition and self-regulation (Nguyen et al., 2023). The phenomenon of "hallucination" in AI tools, wherein they provide scientifically plausible yet factually inaccurate information, is another concern, particularly in the context of GenAI (Surameery & Shakor, 2023). Hence, it is imperative for both teachers and students to critically evaluate the accuracy of generated content and simultaneously develop and continuously improve the skills to create and deliver effective prompts (Lo, 2023).

The historical focus of AIEd research has predominantly centred on technological design and development, driven by researchers with engineering backgrounds (Chiu et al., 2023). However, AIEd is a complex research field that spans across technological, pedagogical, social, cultural, and economic dimensions (Castaneda & Selwyn, 2018). Therefore, AIEd research is scattered and disorganized (Chiu et al., 2023). In general, Miao et al. (2021) argue that the true impact of AI on education remains unclear, making more comprehensive research necessary.

Perception of University Students of GenAI

In addition to the general need of more comprehensive AIEd research, there is a lack of research that considers the perspectives of teachers and students regarding their perception and usage of AI in education (e.g. Chiu & Chai, 2020; Chiu et al., 2023; Hwang et al., 2020).

Kshirsagar et al. (2022) examine human intelligence in the context of AIEd. The research explores how AI is perceived and integrated into teaching and learning. It offers insights into how individuals perceive and utilize AI technologies in educational settings to enhance human intelligence and learning experiences. The authors emphasize the importance of educators and students familiarizing themselves with AI technology to enhance educational abilities, improve outcomes, and make AI education more accessible.

Yüzbaşıoğlu (2021) explores how dental students view and perceive AI in the field of dentistry. About 48 per cent of the 1,103 participating students had basic knowledge about AI technologies and roughly 11 per cent stated that they had no information sources about AI. Approximately 86 per cent agreed that dentistry would be revolutionized by AI but around 29 per cent did not agree that AI could replace itself soon. In addition, roughly 80 per cent of students agreed on topics about AI to be included in undergraduate and postgraduate dental education.

The study by Jha et al. (2022) aims to assess the awareness and views of Nepalese medical students regarding AI in healthcare. In total, 216 students participated. The median AI knowledge score was 11 with a maximum possible score of 25. The score was higher among final year students ($p = 0.006$) and among those with additional training in AI ($p = 0.040$). Half of the students agreed that AI will reduce the number of jobs for doctors. Many expect AI to impact their specialty choice, felt the Nepalese healthcare system is ill-equipped to deal with the challenges of AI, and demanded every student of medicine should receive training on AI competencies.

In their conceptual paper, Baidoo-Anu and Ansah (2023) highlight the benefits and drawbacks of using AI in education, particularly referring to ChatGPT. The benefits are personalized tutoring, language translation, as well as interactive and adaptive learning. On the other hand, ChatGPT has some disadvantages such as lack of human interaction, limited capacity to explain to students, lack of contextual understanding, lack of creativity and privacy concerns by students. Ofosu-Ampong et al. (2023) determined lack of awareness and the disruption of the traditional teacher–student relationship as the main trust concerns. They also highlight that in their survey with 102 participants 20 per cent of students express privacy concerns regarding AI.

By means of an online survey, Kim et al. (2020) show that perceived ease of communication with an AI teaching assistant positively predicts perceived usefulness of an AI teaching assistant. Both perceived usefulness of an AI teaching assistant and perceived ease of communication with an AI teaching assistant show positive relationships with users' attitudes toward using an AI teaching assistant. Attitudes towards using an AI teaching assistant ($B = 0.50$, $p < 0.001$) positively lead to intention to adopt AI teaching assistant-based education. Perceived usefulness positively and directly predicts intention to adopt AI teaching assistant-based education.

The literature review regarding the AI perception and usage of university students shows that most of the studies were conducted with medical students. Hence, research lacks studies in other fields such as the field of management. This research aims to close this gap by examining the AI perception and usage of university students in the field of management.

Consequently, the following research question was derived: *What are the specific perceptions and patterns of usage of GenAI among university students in management?*

Methodology

Between July 6 and August 17, 2023, a qualitative study involving nine bachelor students was conducted through a sequence of interviews. Participants were selected through purposive sampling, ensuring that the selected students have some experience using GenAI and can provide detailed and relevant information regarding the topic.

The interviews lasted between 20 and 45 minutes. The authors attempted to recruit students across a variety of management programmes and semesters. Also, their level of expertise in using GenAI differs. This gave us the opportunity to capture different interviewee perspectives. Table 13.1 presents a summary of the characteristics of the interviewees.

All interviews were conducted through virtual platforms to ensure accessibility and participant convenience. This also allowed for recording and transcription of each interview. To generate insightful qualitative data concerning the participants' understanding, motivations, and challenges of using GenAI software, the interviews focused on the perceived benefits and drawbacks of using GenAI in academic work, as well as the immediate reactions and emotions the participants have towards the user experience. To accomplish this, the interviews followed a semi-structured format. The guideline provided a defined outline to guide the conversation and included a short task

Table 13.1 Overview of interviews and interviewee characteristics.

Interviewee	*Gender*	*Interview duration (min)*	*Study programme and semester*
A	F	20	Marketing Management, 1st semester
B	F	25	Marketing Management, 4th semester
C	F	23	Marketing Management, 6th semester
D	F	30	Marketing Management, 6th semester
E	F	24	Communications and PR, 5th semester
F	F	45	Marketing Management, 5th semester
G	F	35	Marketing Management, 5th semester
H	M	22	Tourism Management, 5th semester
I	F	19	Tourism Management, 5th semester

Source: Own data.

Box 13.1 Sections of semi-structured interview guideline.

Introduction to research, interview confidentiality, and agreement to data use

General questions about understanding and usage of AI, as well as specific experiences and challenges with ChatGPT

Task to write a prompt in ChatGPT to structure a paper and output discussion

Details on impact of AI on, e.g., creativity, future goals, and expectation regarding AI use at university incl. advice for other students

Sign-off with demographic information and room for additional comments

Source: Own data.

utilizing ChatGPT. Thereby, it also allowed for spontaneous and elaborative discussion.

The sections of the interview guideline are briefly outlined in Box 13.1. It aims to prompt the participants to share reflections on their experiences with GenAI, in general, and ChatGPT, in particular. The full interview guideline is contained in Appendix A.

To identify patterns and themes within the data, transcriptions, observational notes, and the researchers' reflective notes were the primary data sources for analysis.

The data analysis followed both an interpretive and an inductive approach. That is, the aim was to understand students' perceptions of GenAI, their emotions, and challenges. Particularly, the data were analysed following the Gioia method, which is often used in qualitative research (e.g. Corley & Gioia, 2004). The method is based on the grounded theory work of Glaser and Strauss (2008). Following Locke (2001), the interview transcripts were coded line by line to find related statements and identify first-order concepts, second-order themes, and aggregated dimensions.

First-order concepts summarize the subjective perspective of the interviewees. The researchers derive the first-order concepts by staying as close as possible to the interview transcripts. Thus, in this first-order analysis, the authors used the participants' wordings to name the codes.

In the second-order analysis, themes were defined that helped to describe the observed phenomena. When the second-order themes emerged, they were assembled into aggregated dimensions (Gioia et al., 2013). Following Gioia et al. (2013), the data structure not only allows for visualization but also shows the progress from raw data to broader themes that indicate rigour in qualitative research (Tracy, 2010).

To improve the data structure, ongoing critical feedback sessions with all authors were an essential part of this research. These feedback loops are a crucial part of the Gioia method because the codes need to be self-explanatory (Gioia et al., 2013).

Results

The applied methodology resulted in the generation of a data structure, as represented in Figure 13.1. This data structure depicts the process by which theoretical concepts from the raw data obtained during the nine interviews were derived. In the forthcoming section, the findings will be presented in a narrative format, illustrating the key concepts through the inclusion of representative quotations extracted from the interviews.

Lack of Tangibility

The interview data reveals that two dominant themes emerged in terms of tangibility of GenAI: *humanization* and *abstraction*. In this context, humanization means that students see GenAI as a real person, whereas abstraction means that they perceive the system as a technology.

On the one hand, students exhibited a personal connection to GenAI, often referring to it as "him" or "her". Additionally, they expressed uncertainty about the appropriate style of communication with these systems, for example whether to use polite language or not.

> "*Now, if you just give him a text and say can you please filter the following out of it, he does it really well*".
>
> "*So, you don't know, do you just say generate me or are you polite and write can you generate me x, y?*"

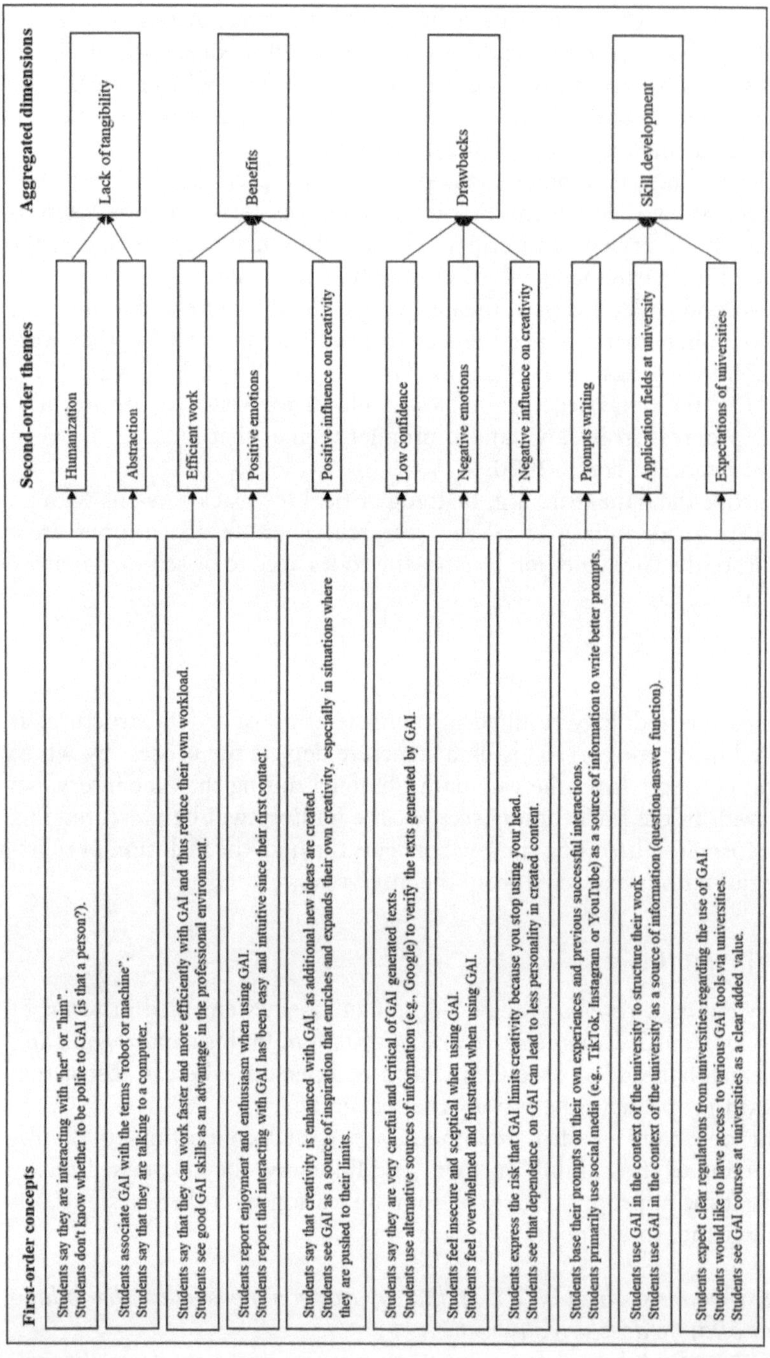

Figure 13.1 Gioia data structure.

Source: Own data.

On the other hand, students commonly associated GenAI with the terms robots, machines, and computers.

> *"I would be too afraid that something completely wrong would be said in my name by this machine"*.
>
> *"That's another weird feeling because you're actually talking to a computer"*.

This finding indicates that students view GenAI as possessing both human-like and machine-like qualities. Thus, GenAI lacks tangibility in the perception of students.

Benefits

Based on our data, three key themes emerged regarding the benefits of using GenAI: efficient work, positive emotions, and a potential positive influence on creativity.

Students highlighted that GenAI facilitated faster and more efficient work processes, providing them with a competitive advantage in their professional environments.

> *"Especially because I do not only content creation at work, but also HTML, CSS . . . it helps me in a fast way to find the right code, which I would not have found faster"*.
>
> *"But the fact that it brings a lot of positive benefits to my job, from my point of view, made me feel that I had to at least deal with it, regardless of whether I personally think it's good or bad, just to stay on the ball for my job"*.

Additionally, students reported experiencing positive emotions such as enjoyment and enthusiasm while interacting with GenAI, emphasizing the ease and intuitiveness of these interactions right from the start.

> *"So more like enthusiasm and euphoria that it worked so nicely"*.
>
> *"So that impressed me the most when I first used it, that it was implemented directly"*.

Moreover, students recognized the potential of GenAI to positively impact creativity, enabling the generation of new ideas and assisting in overcoming creative barriers.

> *"That I might also get ideas through ChatGPT or maybe information that you didn't even think of in the first moment"*.

> *"For myself I think I can rather enrich myself, because I use it so far and also in the future rather in such a way, if I am at my wit's end, to be inspired by ChatGPT and to build on my own ideas".*

In summary, our findings suggest that GenAI offers benefits in terms of efficiency, positive emotions, and the potential to enhance creativity.

Drawbacks

Our data reveal three major themes as drawbacks associated with GenAI: low confidence, negative emotions, and a potential negative impact on creativity.

Students expressed caution and scepticism when it came to the content generated by these tools, often resorting to external sources like Google to verify the accuracy of the output.

> *"However, I would always be cautious with this because it involves information that needs to be technically correct, and I would not trust an AI 100%".*
>
> *"Simply with Google . . . depending on which sources come out there, I click through a bit and if that is already reproduced then on several pages, then I know ok, it seems to be true".*

In contrast to the above-mentioned positive emotions, students also reported experiencing negative emotions such as insecurity and frustration while using GenAI tools.

> *"So, I haven't dealt with it that much yet, but yes, I'm still a bit skeptical and think maybe oh, maybe I'd rather not. Not that you do anything wrong".*
>
> *"But I find somehow overwhelmed . . . , by the fact that I did not quite come to terms with it yet. . . . Frustration perhaps".*

Furthermore, students expressed concerns that reliance on these tools may hinder their own creative abilities, as they no longer rely solely on their own cognitive capacities.

> *"I think it's worse for you in that respect, because then you don't have to use your head as much and your creativity gets a bit lost".*
>
> *"So, a bit that personality of texts is broken".*

In summary, our findings indicate that while GenAI offers valuable benefits as described in the previous section, it also presents significant challenges in terms of confidence, emotions, and creativity for users.

Skill Development

Skill development is the next aggregated dimension which emerged from our data. Our data show three main themes concerning the development of skills in GenAI: prompts writing, application fields at university, and expectations of universities.

Regarding prompts writing, students drew from their past experiences and successful interactions, often seeking inspiration from social media platforms.

> "*I've found my own way to make my results the best they can be*".
>
> "*I've also saved a lot of suggestions from Instagram and TikTok when people tell you to "use this extra info*".

Within the university context, GenAI primarily served as a tool for structuring academic work and obtaining specific answers to queries.

> "*I used her to help create an outline for a presentation, sort of what subtopics or topics she would suggest, and what to include in the presentation*".
>
> "*So, for simple questions, I don't know, if I think of our marketing course, you can just write a question in there and you'll get something as an answer*".

Students expressed specific expectations from universities, which included the provision of comprehensive guidelines for using GenAI, access to tools through university licences, and the availability of practical courses on GenAI.

> "*What we have not yet received from the university was somehow a guideline for this. How should we deal with artificial intelligence, to what extent we can use it at all?*"
>
> "*But there are certainly students who would be interested in the license, for example, just to be able to use that*".
>
> "*I think the best input from the university would be to offer courses to show how to use it without being afraid of it or not knowing what to enter*".

In summary, our findings highlight the significance of prompts writing, application fields at universities, and the expectations students have regarding GenAI skill development.

Discussion

The presented research reveals some paradoxes in the perception and usage of GenAI among management students. On the one hand, the students humanize GenAI. This finding is in line with the study of Wang et al. (2022) in which the authors show that the humanization of GenAI helps students to achieve better learning outcomes. On the other hand, they use terms like robots, machines, and computers to describe this technology.

They also outline that GenAI helps them to work faster and more efficiently, however, they do not trust the outputs of GenAI. They are enthusiastic and enjoy the use of GenAI, at the same time some express insecurity, scepticism, and frustration, as well.

All these contradictions may be caused by the newness of this technology, that is the limited experience of university students with GenAI and a lack of general understanding on how GenAI works. These findings enhance the studies of Kshirsagar et al. (2022), Baidoo-Anu and Ansah (2023) and Ofosu-Ampong et al. (2023). However, interestingly, this research does not find any evidence for concerns regarding data privacy or ethical considerations among management students.

Moreover, the findings show that universities must support students by enhancing their general understanding of GenAI, and their prompt writing skills through dedicated courses on GenAI. This supports the studies of Yüzbaşıoğlu (2021) and Jha et al. (2022) that also outlined the students' need of education regarding AI competencies.

Furthermore, it is essential to establish comprehensive GenAI guidelines for students within the academic context, offering clear instructions on permissible and prohibited uses of GenAI. Crawford et al. (2023) add up on this proposition by suggesting that universities and teachers should lead students to use ChatGPT in an ethical way. The authors posit that instead of frightening students about cheating with ChatGPT, students should be encouraged to have a deeper learning experience with GenAI.

In relation to the impact of GenAI on creativity, the study also shows different perspectives. The students claim that their personal creativity can be enhanced with GenAI, but at the same time they note that GenAI can also limit or even reduce their creativity. Vicente-Yagüe-Jara et al. (2023) show that GenAI supports students' creativity in a language learning setting. They also posit that GenAI cannot replace human intelligence and creativity. Qadir (2023) proposes that GenAI helps students to generate new ideas and hence, foster their creativity. Overall, the discussion in the current AIEd literature is rather positive when it comes to its impact on students' creativity. The findings of this research, however, also show that some students perceive GenAI as a limitation to their own imagination and creativity.

Future Research Agenda

The interpretive and inductive approach of this study offers researchers various starting points for future research, especially considering the above-mentioned ambiguities. The following research questions can help to guide future endeavours in this field:

RQ1: How can universities be motivated to support the students' usage of GenAI?

This research showed that almost all students use GenAI and that students expect support from universities regarding GenAI usage, that is this question aims to explore strategies or incentives that could encourage universities to invest in and support GenAI usage among students.

RQ2: How should a university guideline for GenAI usage look like, that is what are the key elements to be considered?

The study showed that university students expect a clear guideline regarding the usage of GenAI in the academic context. As most of the universities still lack such guidelines, future research could focus on developing a framework with key elements that need to be addressed and considered.

RQ3: What are barriers to fully integrate GenAI in universities?

This question seeks to identify the key challenges and obstacles that limit the deployment and utilization of GenAI in academic settings, despite the notable benefits GenAI offers in the educational scenarios of administration, instruction, and learning.

RQ4: What are the practical applications of GenAI in university teaching?

This question focuses on emphasizing the potential practical applications of GenAI in university teaching, focusing on their capacity to optimize learning experiences for students.

RQ5: What are the factors influencing enhancement or limitation of personal creativity when using GenAI?

This study revealed contrasting perspectives on the potential impact of GenAI on creativity. To gain a deeper understanding, further investigation into potential influencing factors is crucial, such as exploring the possible role of students' general values and attitudes.

In conclusion, this chapter examines university students' perceptions and usage of GenAI within a university context. The explorative study specifically discusses four themes that emerged in the data analysis: lack of tangibility, benefits, drawbacks, and skill development. Based on these findings the chapter proposed five further research questions that can be addressed by future research.

The study faces some limitations, starting with the small interview sample from a single institution and a limited generalizability of qualitative research. In a follow-up study, researchers could conduct more interviews with students from a more varied sample of study programmes and levels, such as Master or PhD programmes. Also, the qualitative results can be tested and verified by quantitative research designs. Overall, this research aims at motivating both universities and their students to critically reflect on their GenAI perception and usage.

References

Agrawal, A., Gans, J., & Goldfarb, A. (2022). ChatGPT and how AI disrupts industries. *Harvard Business Review*. https://hbr.org/2022/12/chatgpt-and-how-ai-disrupts-industries

Baidoo-Anu, D., & Ansah, L. O. (2023). Education in the era of generative artificial intelligence (AI): Understanding the potential benefits of ChatGPT in promoting teaching and learning. *Journal of AI*, *7*(1), 52–62.

Castaneda, L., & Selwyn, N. (2018). More than tools? Making sense of the ongoing digitizations of higher education. *International Journal of Educational Technology in Higher Education*, *15*(22). https://doi.org/10.1186/s41239-018-0109-y

Chen, L., Chen, P., & Lin, Z. (2020). Artificial intelligence in education: A review. *IEEE Access*, *8*, 75264–75278. https://ieeexplore.ieee.org/document/9069875

Chen, X., Zou, D., Xie, H., Cheng, G., & Liu, C. (2022). Two decades of artificial intelligence in education: Contributors, collaborations, research topics, challenges, and future directions. *Educational Technology & Society*, *25*(1), 28–47. www.jstor.org/stable/48647028

Chiu, T. K. F., & Chai, C.-S. (2020). Sustainable curriculum planning for artificial intelligence education: A self-determination theory perspective. *Sustainability*, *12*(14). https://doi.org/10.3390/su12145568

Chiu, T. K. F., Xia, Q., Zhou, X., Chai, C., & Cheng, M. (2023). Systematic literature review on opportunities, challenges, and future research recommendations of artificial intelligence in education. *Computers and Education: Artificial Intelligence*, *4*(3). https://doi.org/10.1016/j.caeai.2022.100118

Corley, K. G., & Gioia, D. A. (2004). Identity ambiguity and change in the wake of a corporate spin-off. *Administrative Science Quarterly*, *49*(2), 173–208.

Crawford, J., Cowling, M., & Allen, K. A. (2023). Leadership is needed for ethical ChatGPT: Character, assessment, and learning using artificial intelligence (AI). *Journal of University Teaching & Learning Practice*, *20*(3), 2.

Gioia, D. A., Corley, K. G., & Hamilton, A. L. (2013). Seeking qualitative rigor in inductive research: Notes on the Gioia methodology. *Organizational Research Methods*, *16*(1), 15–31.

Glaser, B. G., & Strauss, A. L. (2008). *The discovery of grounded theory: Strategies for qualitative research* (3rd ed.). Transaction Publishers.

Hsu, Y.-C., & Ching, Y.-H. (2023). Generative artificial intelligence in education, part one: The dynamic frontier. *TechTrends*, *67*, 603–607. https://doi.org/10.1007/s11528-023-00863-9

Hu, K. (2023). ChatGPT sets record for fastest-growing user base – analyst note. *Reuters*. www.reuters.com/technology/chatgpt-sets-record-fastest-growing-user-base-analyst-note-2023-02-01/

Hwang, G.-J., Xie, H., Wah, B. W., & Gašević, D. (2020). Vision, challenges, roles and research issues of artificial intelligence in education. *Computers and Education: Artificial Intelligence*, *1*. https://doi.org/10.1016/j.caeai.2020.100001

Jha, N., Shankar, P. R., Al-Betar, M. A., Mukhia, R., Hada, K., & Palaian, S. (2022). Undergraduate medical Students' and Interns' knowledge and perception of artificial intelligence in medicine. *Advances in Medical Education and Practice*, 927–937.

Kim, J., Merrill, K., Xu, K., & Sellnow, D. D. (2020). My teacher is a machine: Understanding students' perceptions of AI teaching assistants in online education. *International Journal of Human – Computer Interaction*, *36*(20), 1902–1911.

Kshirsagar, P. R., Jagannadham, D. B. V., Alqahtani, H., Noorulhasan Naveed, Q., Islam, S., Thangamani, M., & Dejene, M. (2022). Human intelligence analysis through perception of AI in teaching and learning. *Computational Intelligence and Neuroscience*, *2022*.

Lo, L. S. (2023). The CLEAR path: A framework for enhancing information literacy through prompt engineering. *The Journal of Academic Librarianship*, *49*(4), 102720. https://doi.org/10.1016/j.acalib.2023.102720

Locke, K. (2001). *Grounded theory in management research*. Sage.
McKinsey. (2023). *What is generative AI?* www.mckinsey.com/featured-insights/mckinsey-explainers/what-is-generative-ai
Miao, F., Holmes, W., Hui, Z., & Ronghuai, H. (2021). *AI and education: A guidance for policymakers*. UNESCO Publishing.
Nguyen, A., Ngo, H. N., Hong, Y., Dang, B., & Thi Nguyen, B.-P. (2023). Ethical principles for artificial intelligence in education. *Education and Information Technologies*, *28*, 4221–4241. https://doi.org/10.1007/s10639-022-11316-w
Nilsson, N. J. (2014). *Principles of artificial intelligence*. Morgan Kaufmann.
Ofosu-Ampong, K., Acheampong, B., Kevor, M. O., & Amankwah-Sarfo, F. (2023). Acceptance of artificial intelligence (ChatGPT) in education: Trust, innovativeness and psychological need of students. *Information and Knowledge Management*, *13*(4), 37–47.
O'Shea, T., & Self, J. (1986). *Learning and teaching with computers: The artificial intelligence revolution*. Prentice Hall Professional Technical Reference.
Ouyang, F., & Jiao, P. (2021). Artificial intelligence in education: The three paradigms. *Computers and Education: Artificial Intelligence*, *2*. https://doi.org/10.1016/j.caeai.2021.100020
Qadir, J. (2023, May). Engineering education in the era of ChatGPT: Promise and pitfalls of generative AI for education. In *2023 IEEE global engineering education conference (EDUCON)* (pp. 1–9). IEEE.
Roll, I., & Wylie, R. (2016). Evolution and revolution in artificial intelligence in education. *International Journal of Artificial Intelligence in Education*, *26*(2), 582–599. https://doi.org/10.1007/s40593-016-0110-3
Sharma, R. C., Kawachi, P., & Bozkurt, A. (2019). The landscape of artificial intelligence in open, online and distance education: Promises and concerns. *Asian Journal of Distance Education*, *14*(2), 1–2. www.asianjde.com/ojs/index.php/AsianJDE/article/view/432
Shi, Z. Z., & Zheng, N. N. (2006). Progress and challenge of artificial intelligence. *Journal of Computer Science and Technology*, *21*(5), 810–822.
Snider, S. (2023). Experts ponder GenAI's unprecedented growth and future. *InformationWeek*. www.informationweek.com/big-data/experts-ponder-genai-s-unprecedented-growth-and-future
Surameery, N. M. S., & Shakor, M. Y. (2023). Use ChatGPT to solve programming bugs. *International Journal of Information Technology & Computer Engineering (IJITC)*, *3*(1), 17–22. http://journal.hmjournals.com/index.php/IJITC/article/view/1679/1993
Tracy, S. J. (2010). Qualitative quality: Eight "big-tent" criteria for excellent qualitative research. *Qualitative Inquiry*, *16*(10), 837–851.
United Nations Education Scientific and Cultural Organization. (2019). *How can artificial intelligence enhance education?* www.unesco.org/en/articles/how-can-artificial-intelligence-enhance-education
Vicente-Yagüe-Jara, M. I., López-Martínez, O., Navarro-Navarro, V., & Cuéllar-Santiago, F. (2023). Writing, creativity, and artificial intelligence. *ChatGPT in the University Context. Comunicar*, *31*(77), 45–54.
Wang, X., Pang, H., Wallace, M. P., Wang, Q., & Chen, W. (2022). Learners' perceived AI presences in AI-supported language learning: A study of AI as a humanized agent from community of inquiry. *Computer Assisted Language Learning*, 1–27.
Yüzbaşıoğlu, E. (2021). Attitudes and perceptions of dental students towards artificial intelligence. *Journal of Dental Education*, *85*(1), 60–68.
Zawacki-Richter, O., Marín, V. I., Bond, M., & Gouverneur, F. (2019). Systematic review of research on artificial intelligence applications in higher education – where are the educators? *International Journal of Educational Technology in Higher Education*, *16*(39). https://doi.org/10.1186/s41239-019-0171-0

Appendix A: Interview structure for study

Introduction

Thank you for taking the time to complete our project. The goal is to understand how students use generative artificial intelligence (GenAI) (such as ChatGPT) in their studies. In this context, we are writing a book contribution for an edited volume that focuses on "Artificial Intelligence, Co-Creation and Creativity". You can always skip a question if you don't want to or can't answer it.

Is it okay if I record this interview so I can transcribe it for our research? I will keep all data confidential, and any research findings will be presented anonymously.

Before I start recording, do you have any questions about this interview?

Warm-up questions

What does artificial intelligence (AI) mean to you or how would you define it in your own words?

What AI tools are you familiar with? What AI tools have you used before? Do you have a favourite? If so, why?

How long have you been actively using AI tools?

How often do you use AI tools?

Motivation for using GenAI

Let's take ChatGPT as an example. In what situations or for what purposes do you use ChatGPT in your studies?

Do you find it easy to interact with ChatGPT?

To what extent do you find ChatGPT useful for your studies?

Advantages of using ChatGPT

What do you see as the core benefits of using ChatGPT in your studies?
Can you give me a specific example of how it facilitates your studies?

Challenges in using ChatGPT

Where do you see the biggest difficulties in using ChatGPT in your studies?
Have you ever thought about stopping using ChatGPT? If yes, why?
What emotions do you feel when you use ChatGPT? Can you give me an example?
How do you find the appropriate requests or prompts for ChatGPT?

ChatGPT task

Now I would like to see how you use ChatGPT. For this, I share my screen and ask you to formulate a prompt (a request) for the following situation: Assignment – you want to write a term paper/project work on the topic "Using generative artificial intelligence in marketing".
How do you use ChatGPT to get started with this term paper/project work? What prompt (what request) are you using?
Let's take a look at ChatGPT's response together. Does it meet your expectations? Is it helpful? Why (not)?
What do you do in the next step?

Outlook

Do you have future goals regarding your GenAI use in your studies?
How can the university help you make better use of GenAI?
Have you specifically trained or helped others to use GenAI?
What advice would you give to other students who want to use GenAI?
What impact has GenAI had on their personal creativity, or do you think your personal creativity will change because of this?

Demographic variables

Gender, Study Programme, and Semester

Additional comments

Do you have any additional comments on this topic?
Thank you very much for the interview!

Source: Own data.

Chapter 14

Co-Creation with Artificial Intelligence in Marketing

The Case of the Data Analytics Company Dataiku

Nicole Lontzek

Introduction

The recent developments in the field of generative artificial intelligence (AI) have reignited discussions regarding its application within the field of marketing. Within the contemporary landscape, a marketer finds oneself encompassed by a wide range of AI-supported tools, all designed with the aim of simplifying people's lives, enhancing creative capabilities, and augmenting efficiency and effectiveness. While there are various potential use cases, this chapter will delve into a specific B2B marketing scenario within the software industry.

Dataiku is a globally operating software provider for AI and data analytics, which was founded in Paris in 2013 and is now headquartered in New York City. The company employs over 1,200 employees with offices around the world. Dataiku's annual recurring revenue passed the $230 million mark in 2023 (Douetteau, 2023).

The company serves customers across industries, such as the semiconductor company NXP, the consumer goods producer Unilever, the automotive manufacturer Mercedes Benz, the pharma giant Roche, and the reinsurer Munich Re.

The company's software solution enables data experts and domain experts alike to work within one central platform and create their own machine learning models, independently of one's prior data science or AI expertise. With that approach, Dataiku democratizes the use of data and machine learning models by reducing complexity and making the technology accessible to everyone.

Additionally to the proprietary platform, the company provides a wide range of services – from implementation to long-term strategic planning. The software processes data from various sources and entails a library of predefined machine learning models and the option to create proprietary models depending on the use case. The tool provides recommendations, simulations, predictions, and visualization.

DOI: 10.4324/9781003453901-18

The core message of Dataiku revolves around "Everyday AI, extraordinary people" and marketing addresses two personas. One is focused on decision-makers in the AI and data field, such as "Chief Digital Officer" and the other one addresses practitioners, such as "data scientists". The core pain points of the personas are: a missing AI strategy to begin with, a lack of transparency in data usage, silo-thinking and working in the different departments, a lack of trained staff, or too many daily mundane repetitive tasks, which prevent the teams from working on innovative forward thinking projects.

Dataiku addresses these pain points with its software solution and services and promotes its offering with a series of marketing campaigns throughout the year.

The marketing department employs an all-round approach, using various tactics and channels, ranging from content creation, performance marketing, partner marketing to events and thought leadership practices. Dataiku's´ marketing campaign structure is divided into brand campaigns and industry vertical specific campaigns. Campaigns run globally and on a local level to accommodate the specific needs of each market. A campaign typically runs for the duration of six months.

Each region has an appointed local marketing director, who steers the activities and has full ownership of it. The local marketing directors align with the internal marketing sub-functions, such as marketing analytics, paid search, partner marketing, social media, content marketing, and design to execute the global marketing strategy and localize it to meet specific needs for their markets.

With that in mind, the local marketing directors decide on which campaigns should be localized. They take cultural aspects of the markets into account by translating the various assets into the regional language while respecting local market behaviour, preferences, and the tone of voice.

Success criteria for marketing activities are revenue generation and brand awareness. Revenue generation is measured by the progression of marketing qualified leads (MQL) that convert into pipeline growth. Brand awareness is monitored by the "share of voice" in the markets achieved through PR activities.

While the proportion of companies adopting AI has plateaued, the companies that have adopted AI continue to pull ahead, and recent breakthroughs within the field of generative AI, in 2023, sparked new demand from existing clients and prospects, causing Dataiku to pivot its strategy and extend its offer to generative AI capabilities within its proprietary software platform, Data Science Studio (DSS) (AI Index Report 2023 – Artificial Intelligence Index, O. D.) (Dataiku, 2023). Therefore, new functions, including large language model (LLM) application programming interfaces, have been developed and released in June 2023.

In response to the newly developed functionalities, the marketing team promptly initiated an *ad-hoc* campaign focused on "generative artificial intelligence" to capitalize on the prevailing momentum. Given the urgency and relevance of the campaign, the team needed to shorten its launch time for the campaign from typically four months to two months, while leveraging the existing teams' capabilities and budget by using AI-supported tools.

Dataiku's marketing campaigns consist of an array of content assets, including E-books, Whitepapers, blog posts, PR outreach, and a series of targeted emails. Complementing these efforts, a performance marketing campaign was created, featuring a diverse range of text and creative ads for search engines and social media.

One previous campaign, which was launched in early 2023, was dedicated to the topic of upskilling one's workforce for AI and machine learning applications. The campaign's name was: "Upskilling – How to Win the Battle for AI and Data Talent". The target audience of this campaign was AI – and data leaders, who were facing the challenge of skill shortage in the AI space. This campaign addressed how to overcome skill shortage by upskilling other domain experts to be data workers. The campaign also discussed organizational structure fit for supporting upskilling initiatives. It addressed the scarce and underutilized data workers' pain. This challenge is commonly experienced by data science and analytics leadership and often results in loss of productivity and talent churn. The campaign was rolled out across all industries.

The bill of assets included a wide range of content from E-Books (*Upskilling: How to Win the Battle for AI – and Data Talent*) (Dataiku, o. D.), customer stories from pharma giant Pfizer on empowering a global analytics team with the right skills, from the British multinational insurance company Aviva on how they navigate the data talent shortage (Dataiku Community, 2022) videos and blog posts ("Align your operating model with your AI maturity") (Bryan, 2023). The campaign also included a series of performance marketing campaigns in the form of paid ads and social media outreach. It was rolled out globally and localized for some markets: DACH, Japan, and France. The campaign development process for the global campaign took the team four months from November 2022 to February 2023. The localization took an additional month since the regional directors are dependent on the production of the global content assets.

Since each campaign creation and localization process is time-consuming and resource-intensive, the marketing team changed its approach for the generative AI campaign and explored the collaborative potential of AI-supported tools for their own marketing efforts and found many opportunities to accelerate and enhance the campaign development process.

Each campaign consisted of an overarching topic and several subtopics to address the personas' pain points. Therefore, after formulating the primary message, the task of delineating subtopics commenced.

At this juncture, AI assumes a pivotal role in the campaign creation process by participating in the research and ideation phase to generate subtopics of interest for the target audience.

After the research and ideation is completed, the localization and content generation follows, as well as the creation of performance campaigns and the monitoring of it. In all stages of this process, AI served as a valuable co-creator.

It is worth noting that these tools operate on the foundation of extensive datasets and are developed with predefined rule-based actions. Consequently, the user must ensure that the tool prompt is highly specific to yield desirable results.

It is crucial, however, to elucidate that these tools are susceptible to errors, necessitating validation of their generated output through reputable sources and quality checks. Additionally, it is essential to acknowledge their limitations, notably the absence of access to real-time data sources, such as web browsing capabilities, which inherently constrains their functionalities. Consequently, an integral facet of this process entails a quality assessment to authenticate the outputs, underscored by the indispensable role of human judgement. Within the continuum of the campaign creation process, the synergy between the marketer and the AI entity manifests as a collaborative endeavour, characterized by a harmonious interplay between their respective strengths.

Discussion

Comprehensive marketing campaigns with AI-supported tools as co-creators for a B2B software company adapting to changing market conditions.

In the pursuit of comprehensive marketing campaigns, with an emphasis on adaptability to shifting market conditions, Dataiku endeavours to extend its marketing reach and localize campaigns with greater speed and precision.

The marketing team operates within a well-defined structure, led by regional marketing directors. These directors play a critical role in accommodating the unique characteristics of various local markets, including but not limited to APAC, EMEA, North America, South America, and the Middle East. They address the nuanced differences within individual countries to effectively tailor their strategies.

Market dynamics differ, with the US market showing greater receptivity towards the company's branding initiatives, while the German market leans towards relying on empirical data and factual insights to guide their decision-making processes.

The regional marketing directors ensure they consider these differences within the localization process, while keeping the central message woven into the campaign. A comprehensive marketing content calendar serves as the bedrock of the overarching marketing strategy. This calendar encompasses a spectrum of campaign components.

Crucially, in light of the ever-evolving market dynamics, the company recognizes the indispensability of adaptability, necessitating the vigilant monitoring of market trends and the swift execution of supplementary (local) marketing campaigns that may extend beyond the predetermined scope.

The developments in the AI domain in 2023 have thrust Dataiku into the forefront of industry discourse. In response, the company has pivoted its software offerings to align with the evolving standards, emphasizing generative AI capabilities.

The marketing team, in turn, confronts the challenge of rapid adaptation to this paradigm shift, mandating operational excellence in crafting an *ad-hoc* marketing campaign covering the latest developments and crafting localized campaigns tailored to the company's core markets. It is within this context that the concept of co-creating with AI assumes a pivotal role.

Research and Ideation

The planning of an *ad-hoc* marketing campaign centred on generative AI demands, quickly understanding the market's needs, translating them into valuable resources, and close cross-department collaboration. Therefore, the regional marketing directors utilized three tools for co-creation with AI in the research phase, gong.io (Gong, 2023), otter.ai (Otter.ai – AI meeting note taker and real-time AI transcription, o. D.-b), and Dataiku DSS.

The directors concentrated on three key performance indicators (KPIs). These metrics encompassed the hours allocated to the research phase using leveraging tools, the quantity of team members engaged in this process, and the precision of the generated outcomes.

This precision entailed both voice-to-text protocol translations and the analysis of text inputs via crawlers to identify the most pertinent topics. The research phase entails analysing recorded customer calls, marketing surveys for existing clients, and close collaboration with Dataiku's partner ecosystem with companies like Deloitte, Snowflake, and Databricks. Additionally, it demands engaging with analysts from BARC (König, 2023), Forrester (Forrester, 2023), and Gartner (Gartner | Delivering actionable, objective insight to executives and their teams, o. D.) and analysing their market reports.

These insights are documented in Dataiku's customer relationship management (CRM), in the partner portal and web analytics tools and streamlined and processed via LLM within Dataiku's proprietary AI software, DSS. The sales calls are recorded within the sales software gong.io and transcribed with otter.ai.

Streamlining these data enables the marketing team to identify customer needs quickly through analysing recorded and transcribed sales calls by searching for pain points and keywords within DSS and by performing sentiment analysis of survey feedback, analyst reports, and transcribed calls. This allows the marketing team to automatically crawl through large proprietary datasets in a friction of the time of what a human colleague could achieve. Identifying these customer needs is crucial for the campaign's subtopics.

To facilitate this, the marketing analytics team partnered with the data science team to create an LLM dedicated to AI-supported research in DSS. After understanding the requirements and successfully creating a minimum viable product (MVP), the data science team rolled out the LLM, visualized the results in the DSS dashboard, and made them accessible to the marketing directors, who identified two major subtopics to focus on, since they were frequently mentioned, in the analysed surveys, calls, and reports, which were "responsible AI" and enterprises getting started with generative AI, missing an orientation point and guidance.

Subsequently, the ideation phase ensues, necessitating the synthesis of the insights and conclusions gleaned from prior research efforts, thereby contextualizing the implications for the intended target audience. In the research and ideation phases, AI emerges as a trustful co-creator, offering support in navigating the complexities of the contemporary marketing landscape.

Its capacity to process vast datasets, discern patterns, and generate insights positions it as an indispensable ally in the pursuit of crafting campaigns that resonate with the audience. AI tools help synthesize vast and diverse information sources to provide a comprehensive understanding of market needs. These tools can analyse and summarize various forms of data, including human-generated meeting protocols, recorded sales calls with customers and prospects, as well as publicly available market reports crafted by analysts.

These insights were integrated into the ideation process to develop content in form of E-Books (*The C-Suite Guide to an LLM strategy*) and a series of blog posts, such as "How to go from POC to LLM in production" (Heidmann, 2023). To facilitate the team's efforts during the ideation phase, local marketing directors utilized ChatGPT-4 and HubSpot's AI content writer (Free AI Content Writer | HubSpot, o. D.). The marketers first used ChatGPT-4 and customized prompts. considering local nuances, specifically targeting the themes of responsible AI and getting started with generative AI.

One such prompt for France reads: "*What are the main concerns regarding 'responsible AI' in France?*" This command prompts the tool to generate a list of bullet points outlining key considerations for the respective markets, such as algorithmic transparency and biases, ethical and legal frameworks, and data protection and privacy.

While these bullet points offer initial direction and support, it's crucial to internally validate their accuracy with subject matter experts and clients to ensure precision. Another prompt dedicated to Germany reads: "What are the specific data privacy regulations for the AI in Germany?"

Additionally to using ChatGPT-4, the marketers utilized HubSpot's AI content writer, which was released in March 2023 to generate copy ideas and variants for content on the topics "responsible AI" and "Getting started with generative AI". The tool relies on human input, such as the desired topic, tone, length of the copy, and reading level. Overall, the marketing team requires up to two weeks for the research and ideation phase. Utilizing Dataiku's DSS, otter.ai, ChatGPT-4, and HubSpot's AI content writer, the team significantly expedited this phase, reducing it to five business days. This achievement not only set a new standard within the company, but it also ensured precision in results while eliminating guesswork.

While data-driven insights provide clear implications regarding market needs, it is the prerogative of the marketer to discern which topics warrant specific attention within the campaign's purview. This structured approach aims to engender broad-reaching and highly relevant campaign content for the target audience.

Localization and Content Generation

Cultural disparities and regional preferences significantly influence consumer behaviour and demand a nuanced approach. Hence the regional marketing directors of Dataiku tailored the content of the generative AI marketing campaign to suit the specific nuances of Japan, France, and the DACH region. To ensure these assets resonate with the specific markets, presenting the content in the local language is important.

Translating expert content accurately while preserving local expressions can prove to be a time-intensive endeavour. In this regard LLMs offer invaluable support to the marketers. These tools expedite the localization process, as marketers can submit content with a prompt to translate it into the desired language and verify the output based on human judgement. For the translation of content assets, Dataikus local marketing directors utilized DeepL, a tool from an AI-software company founded in Germany in 2017.

The tool supports the translation of 31 languages in 2023 (DEEPL Übersetzer: Der präziseste Übersetzer der Welt, o. D.). This approach enables the translation of content pieces in a fraction of the time needed by a human. A typical E-Book published by Dataiku contains an average of 20 pages to be translated and takes a translator seven business days. Leveraging DeepL shortened the time to two days, including quality checks.

However, it is important to note that the efficacy of AI-supported translation varies depending on the target language. These models are primarily trained on widely spoken languages and may have limited data available for less commonly used languages. Additionally, since the generated output originates from a machine, the translated text may feel robotic for the reader. To mitigate these limitations, marketers can employ follow-up prompts to modify the generated texts, but they also need to apply human judgement.

This process empowers marketers to fine-tune the translations and ensure that the content remains culturally sensitive and contextually relevant, thereby bridging the gap between linguistic diversity and effective communication.

AI tools require careful handling and companies should refrain from inputting sensitive company information into these tools, as they utilize such data to enhance their machine learning models, potentially making it accessible to their entire user base. To avoid competitive disadvantages or legal repercussions, individuals should seek guidance from their legal departments.

While the core elements of intellectual property and creativity remain within the purview of humans, AI tools excel in accelerating the localization process. They offer an initial draft of translations, taking cultural nuances and the preferred tone of the target language into account.

In the final phase of content creation, human intervention becomes indispensable. Marketers are tasked with conducting a quality check, ensuring that the translated content not only aligns with linguistic accuracy but also maintains cultural sensitivity and contextual relevance.

Execution Excellence

In the post-content generation phase, the distribution efforts of marketing campaigns through various channels commence within Dataiku. These efforts encompass a combination of email outreach, social media engagement, and performance marketing initiatives. While crafting performance marketing outreach, the creation of text ads for paid search efforts and creatives for complementary social media campaigns is a key aspect for the success of each campaign. In this phase, AI tools can support execution excellence within the time-sensitive marketing environment.

To maintain high-quality text content, marketing directors formulated initial examples of the desired outcome themselves, respecting the constraints of the tools. They provided a variety of text inputs for ads. One text input example reads: "Unlock your data value. Upgrade to Gen AI with Dataiku. Equip your experts with the tools they need to level up your AI game!"

Subsequently, the team leveraged ChatGPT-4 as well as HubSpot's AI Google ad copy generator to generate a diverse set of ideas from these examples, considering the constraints predefined by the team, such as the purpose of the campaign, limited number of characters, the desired language, or tone of voice of the ad copy (Pinon, 2023; HubSpot, o. D.). Co-creation unfolds as the marketer provides input data and AI processes and expands the creative work.

Co-creation with AI tools also encompasses the generation of visual creatives for marketing campaigns and marketers sought to expedite the banner creation process.

AI tools like OpenAI's DALL·E 2 supported the marketers to generate original, realistic images from textual inputs through natural language processing. The local marketers initiated this process by manually creating examples of creative banners, considering central messaging, corporate identity, and the target audience. The input was then fed into DALL·E 2, which swiftly generated creatives based on the input, significantly accelerating and expanding the creative process.

Monitoring

Following the campaign's launch, diligent monitoring of asset performance is the key. AI assumes a central role in this phase, utilizing historical marketing data analysis of content assets from previous campaigns to compare current campaign results. Analytics dashboards, routinely employed for the scrutiny of campaign performance, feature integrated machine learning models designed not only for monitoring purposes but also for optimization recommendations. These recommendations are contingent upon the accumulation of a minimum threshold of requisite data to ensure reliability. A practical example is the monitoring of translated content assets for the localization of the generative AI campaign. Each campaign is documented and structured in Dataikus content management system (CMS) and CRM solution, HubSpot. The tool provides the function to set up each campaign within it, formulating the company's goal, such as lead generation. Within this setup, considering the overarching structure and goal, marketers add all the generated content assets, from global assets to localized ones. The tool monitors the leads generated and the associated potential pipeline for each content asset, cumulates the performance of each asset in the campaign, and thus calculates the overall performance of the campaign. This allows everyone in the marketing team to not only evaluate global campaign performance but also break down performance on a local level. In order to generate sufficient data and make an initial performance assessment, a minimum duration of three months after the start of the campaign has proven to be practicable. Blog posts and E-books were evaluated in HubSpot's CMS and CRM solutions using HubSpot's integrated AI to determine suggestions for future content optimization. The AI analyses the readability of content, time spent on the assets, and bounce rates to increase the performance of the two main KPIs for this matter, which are generated MQL and campaign influence on target accounts, which ultimately result in pipeline generation.

As the campaign is still ongoing while this case is being written, the final results of the campaign need to be awaited. However, the utilization of AI for co-creation allowed the marketing team to accelerate the development process by automatically transcribing and crawling the recorded sales calls, and crawling surveys and reports for the most pertinent topics within the research phase. In the ideation phase, the use of LLMs enabled extended brainstorming possibilities for sharpening the campaign's sub-themes through access to large datasets. Local marketers profited from the *ad-hoc* translation capabilities of AI for localizing content. Enhanced human creativity was also reflected in the execution phase, where marketers provided initial examples of ad copy and visual banners and the AI-generated variations of these. Last, the monitoring of the campaigns success was supported by AI through comparing historical and current marketing campaign data and providing optimization recommendations. In summary, it can be said that throughout the entire campaign process, the interaction between humans and AI enabled the process to be accelerated and human creativity to be expanded.

Future Research Agenda

Given the optimization of processes and the challenges of using AI to develop the marketing campaign described in this chapter, maintaining a consistently high quality of creative content is critical to its success.

RQ1: How can marketers ensure a consistently high standard of co-created creative outputs with the assistance of AI?

AI tools are prone to errors and are largely limited to generally accessible data. Therefore, marketers must ensure that the quality of the co-created outputs meets their quality standards, and they need to set clear guidelines. Although AI significantly supported the campaign development process, it also showed weaknesses. The transcription of the calls contained contextual errors that had to be corrected manually. In the ideation phase, marketers needed to provide accurate initial input, because the tools access general data and results are otherwise diluted. Training the tools to understand the brand's tone, style, and preferences helped in enhancing the relevance of generated content.

Translation capabilities of specialized content already provided good results; however, errors occurred. The more specific the content was, the more errors occurred.

System updates of some tools caused a loss of data and therefore lowered the output quality. Backup data are paramount here. Overall, human oversight in the form of review processes is the key for consistently high quality.

Due to the rapid development of AI and its extension of capabilities, creative professionals are experiencing a shift and reorganization of the distribution of tasks between humans and machines.

RQ2: How will the role of creative professionals change with the rise of AI?

AI is increasingly used to automate repetitive tasks like data analysis and content distribution, allowing creative professionals to focus on strategic and creative aspects (Davenport, 2023). It provides access to analytics for more informed decisions, aiding in understanding customer behaviour for targeted campaigns.

AI-driven personalization tools tailor marketing messages to individual preferences, enhancing conversion rates. AI, such as LLMs, assists in content generation and can inspire new ideas, making it a tool to amplify creativity. Additionally, co-creation with AI enables quicker responses to changing market conditions, demanding agility from creative professionals.

Technical proficiency of the marketer in understanding how AI algorithms function in a marketing environment and data analysis becomes essential for effective collaboration with AI. Traditional silos between roles are breaking down, requiring creative professionals to integrate creative thinking with data-driven insights and technical expertise. The co-creation with AI will redefine the role of creative professionals in marketing by demanding a blend of creative, technical, and analytical skills.

As co-creation with AI continues to spread, the question of how to deal with copyright issues when creating marketing campaigns collaboratively with AI arises.

RQ3: What are the procedures for managing copyright concerning creative work collaboratively generated with AI?

Intellectual property of co-created work is a sensitive topic and needs to be handled carefully. But the question on how to treat the copyright for co-created work with AI still remains unanswered as there is no legal basis in 2023.

Today's laws have not yet been adapted to the use of technologies such as generative AI. The European Union (EU) AI Act is the closest the EU countries have gotten to regulate AI use, specific laws for copyrighting AI-generated content are yet to be established and until then it might just be dealt with on a case-by-case basis (Atillah, 2023). The US law, in contrast, denies copyright for co-created work as this chapter is written, as it was ruled by a US court judge in 2023 (Feed, 2023). As AI continues to rise, legal regulations still need to be defined.

References

AI Index Report 2023 – Artificial Intelligence Index. (o. D.). https://aiindex.stanford.edu/report/

Atillah, I. E. (2023, July 11). Copyright challenges in the age of AI: Who owns AI-generated content? *Euronews*. www.euronews.com/next/2023/07/10/copyright-challenges-in-the-age-of-ai-who-owns-ai-generated-content

Bryan, D. (2023, February 20). *Align your operating model with your AI maturity*. https://blog.dataiku.com/align-your-operating-model-with-your-ai-maturity

Dataiku. (o. D.). *Upskilling: How to win the battle for data + AI talent*. https://pages.dataiku.com/battle-data-ai-talent

Dataiku community. (2022, 15. September). *Dataiku community*. https://community.dataiku.com/t5/Dataiku-Frontrunner-Awards/Aviva-Deploying-a-Standardized-Centralized-and-Modernized/ta-p/28854?_gl=1*1mzwvot*_ga*NTY3NDI0MzQ3LjE2Njk0NjY0NDI.*_ga_B3YXRYMY48*MTcwMDM4OTY4MC4xMi4xLjE3MDAzOTA4MTIuMTEuMC4w

Dataiku. (2023, 7. November). *Dataiku product | everyday AI, extraordinary people*. www.dataiku.com/product/

Davenport, T. H. (2023, August 15). How generative AI is changing creative work. *Harvard Business Review*. https://hbr.org/2022/11/how-generative-ai-is-changing-creative-work

DEEPL Übersetzer: Der präziseste Übersetzer der Welt. (o. D.). www.deepl.com/de/translator

Douetteau, F. (2023, September 6). *We were built for this moment.* https://blog.dataiku.com/we-were-built-for-this-moment

Feed. (2023, 21. August). Artificial intelligence (AI) generated artwork can't be protected by copyright, ruled US court judge. *The Economic Times.* https://economictimes.indiatimes.com/news/international/us/artificial-intelligence-ai-generated-artwork-cant-be-protected-by-copyright-ruled-us-court-judge/articleshow/102918328.cms

Forrester. (2023, Oktober 30). *Forrester helps organizations grow through customer obsession.* www.forrester.com/bold

Free AI Content writer | HubSpot. (o. D.). www.hubspot.com/products/cms/ai-content-writer?hubs_content=www.hubspot.com%2Fproducts%2Fartificial-intelligence&hubs_content-cta=wf-link-to-content-card-text-link

Gartner | Delivering actionable, objective insight to executives and their teams. (o. D.). *Gartner.* www.gartner.com/en

Gong. (2023, November 30). *Request demo | Gong.* www.gong.io/win-with-gong/

Heidmann, L. (2023, November 21). *How to go from POC to LLM in production.* https://blog.dataiku.com/llm-in-production

HubSpot. (o. D.). *AI Google ads copy generator | Hubspot.* www.hubspot.com/campaign-assistant/ai-google-ads-copy-generator

König, H. (2023, Oktober 31). *BARC – data decisions: Built on BARC.* https://barc.com/de/

Otter.ai – AI meeting note taker & real-time AI transcription. (o. D.). https://otter.ai/

Pinon, L. (2023, June 22). ChatGPT for Google Ads: How to enhance campaign creation. *Search Engine Land.* https://searchengineland.com/chatgpt-google-ads-campaign-creation-428533

Chapter 15

Artificial Intelligence as a Creative Scientist

Markus C. Hemmer

Creativity in Research

Artificial intelligence (AI) opens a range of opportunities to support creative processes in research and development. Intuition and creativity are required in several phases of scientific endeavours (for an overview on applications, see Kara, 2020). In the early phase of research, formulating hypotheses that explain observed phenomena or predict outcomes is a creative process. Scientists often need to think creatively to generate hypotheses that push the boundaries of current knowledge.

In the exploratory phase, hypotheses from abductive reasoning serve as the best explanation of phenomena that have not yet been fully investigated. Scientists must then consider alternative explanations and exclude them to improve the validity of their hypotheses. In the descriptive phase, statistical modelling supported by data science and AI approaches requires creativity in the identification of adequate probability distributions and experimental parameters. The explanatory phase takes advantage of creative solutions such as artificial neural networks and other machine learning approaches to obtain reasonable explanations for cause-and-effect relationships to finally arrive at the predictive phase of research.

Throughout the research process, unexpected challenges and roadblocks may arise. Creative problem-solving skills are then essential to devise alternative strategies, adapt methodologies, and optimise parameters for algorithmic or experimental settings.

A creative mind is also required to conceive and develop instruments that enable novel experiments and data collection methods. Much scientific advancements are driven by the invention of innovative technologies, conceptual frameworks, or models that explain complex phenomena. These frameworks often require synthesising existing theories or paradigms in innovative ways. Creative thinking is also vital to identify synergies between disparate scientific areas and apply insights across domains. This requires visualising complex scientific concepts which again involves creative skills

DOI: 10.4324/9781003453901-19

to design informative figures, schemes, and diagrams to communicate and validate research findings effectively.

Among all these demands for creativity in scientific endeavours, the probably most important contribution of AI lies in its ability to identify and explain relationships.

AI's Explanatory Power

Observed or measured phenomena require formulating novel and testable hypotheses. Such hypotheses must be empirically plausible in terms of their relationship to the context conditions and a predictable variable. In addition, they must be interpretable by scientists. Hypotheses are generated by observing patterns or contradictions to related phenomena. Systems comprising supervised learning (learning by example) can predict accurately new data after adequate training. Unsupervised (self-learning) methods allow for identifying patterns and contradictions that are not obvious to the researcher. This applies particularly to the fields of data science and big data, where the information content is often overwhelming to a scientist. Finally, reinforcement learning (learning by response to actions) enables computer systems to react to environmental conditions in real-time scenarios.

A first step in research is often the design of experiments. Designing experiments that effectively test hypotheses while accounting for adequate experimental parameters and effective variables also require creativity. A corresponding AI approach to the design of experiments is given by Lauro et al. (2015). The authors compare artificial neural networks, fuzzy logic, genetic algorithms, and Bayesian approaches to identify adequate model parameters and optimise them according to the desired response variables. A remarkable example of AI-supported systems that cover both formulation and testing of hypotheses in the pharmaceutical field are robot scientists – fully automated robots that generate hypotheses, design and conduct experiments, and interpret the results to eventually adapt the experimental settings or validate the outcomes (King et al., 2023).

Interpreting hypotheses tests requires creativity in data analysis to draw meaningful conclusions. Data science is a particularly creative field where innovative statistical and machine learning approaches help to extract new insights. However, the structure of enormous amounts of data processed in data science is often hard to elaborate by humans. AI supports the creative process by interpreting the structure that underlies large amounts of data. AI can even outperform expert scientists in interpretation, as shown by a study on pulmonary function tests (Topalovic et al., 2019).

AI algorithms can analyse vast amounts of scientific data, identifying patterns, trends, and correlations that may not be immediately apparent to human researchers. By leveraging machine learning techniques, AI can assist in data-driven discoveries and co-create insights by uncovering hidden patterns within complex datasets. An extensive review on technologies and applications of AI-based pattern recognition is provided by Abiodun et al. (2019). For example, supervised, unsupervised, and reinforcement learning are successfully applied to image analysis in radiology for tumour segmentation and treatment optimisation (El Naqa et al., 2020). More sophisticated approaches, like self-supervised learning (learning by self-generated examples) and geometric deep learning (utilising symmetries in data), were used to analyse the structure of scientific data to propose new levels of abstraction that a scientist can use to generate explanatory models (Wang et al., 2023).

Explanations beyond the Accessible

Although these examples might seem limited to immediately measurable data, several approaches were successful in the field of cognitive sciences using qualitative data. For instance, Briganti (2022) presents a method of hypothesis generation with Bayesian network models in psychiatry, and Oquendo et al. (2012) described the implementation of data mining and machine-learning in healthcare to generate hypotheses for pathophysiological models. They used implications from scientific literature for hypothesis generation and identified genetic and neurobiological causes of psychiatric disorders. Even the legal system can take advantage of AI as shown by a working paper of the US National Bureau of Economic Research where AI-based hypothesis generation supports the decision of a judge in legal cases (Ludwig & Mullainathan, 2023).

The ability of AI to generate novel research ideas by analysing existing data can be realised with generative AI models that create new data or information (for a tutorial, see Goodfellow, 2017). Generative models can also help to identify potential collaborators with complementary skills and interests and funding opportunities based on individual researchers' expertise, fostering co-creation and interdisciplinary collaboration (see, for instance, Yuan et al., 2020; Shao et al., 2020). AI can facilitate virtual collaboration platforms where researchers from diverse backgrounds can connect, share ideas, and co-create research projects (Griffith & Mangla, 2023). An interesting review on collaborative AI-based real-time interactive improvisation for music creation is given by McCormack et al. (2020). The authors emphasise the need for systems that can "articulate or explain the decisions they have made" (McCormack et al., 2020, p. 42) and propose a design process for creative practitioner teams. The organisation of such teams (including the AI system) and the role creative artists play by using an AI generative model seem to be particularly important (Mazzone & Elgammal, 2019). The same applies to engineering, where collaboration between teams and AI systems is deemed an important aspect in research, conception, and the modelling phases of engineering design processes (Figoli et al., 2022).

Intelligent Knowledge Review and Synthesis

Another field where AI strongly supports the scientific community is the natural language processing and generation. AI-based systems can aid in conducting comprehensive literature reviews by automatically analysing vast amounts of scientific articles and papers for patterns, similarities, contradictions, and inconsistencies. This alone would promise a dramatic increase in the quality of scientific documentation in the future. Natural language processing helps researchers to identify relevant studies, extract key information, and synthesise knowledge from multiple sources if several conditions to formulate the query are considered (see Wang et al., 2023).

Natural language processing and its ability to assist writing supports (collaborative) scientific documentation by grammar and style suggestions, whereas generative art systems provide the visualisation to improve the communicative value of studies (for a short introduction and ethical issues, see Salvagno et al., 2023). Natural language processing and generation support scientific communication in terms of writing style, translation, grammar, sentence structure, citation, and referencing (Huang & Tan, 2023). Ray (2023) provides a review on ChatGPT and its use in sciences. The author also emphasises how ChatGPT can find latent patterns of data as well as identify gaps and deficiencies in academic publications. Although there are challenges in terms of authorship, inaccuracies, and ethical issues (Xames & Shefa, 2023), as well as open research questions in language modelling and verification (Van Dis et al., 2023), natural language processing already reshaped the process of scientific documentation.

Some scientific disciplines are still in the phase of evaluating the potential and issues of AI supporting research. For instance, Zhai et al. provides a review on the development and research issues in educational sciences in the past decade (Zhai et al., 2021) and emphasise the need for educators to collaborate with AI engineers to efficiently bridge the technology gap. Although the application of creative AI is still in a state of infancy in the social sciences, natural language processing and machine learning capabilities open a series of opportunities. For instance, Grimmer et al. (2021) explicate how a transition from deductive to the desired inductive approaches to inference can be achieved by machine learning and highlight the therefore increasing importance of formal theories in social sciences.

The Way to Scientific Understanding

Although scientists would look forward to AI solutions that can predict all potential explanations to a phenomenon or present all solutions in a logical domain, they would demand an explainable system. Krenn et al. (2022) referred to the key point in such system as "scientific understanding" in the sense of the philosophy of science. The authors distinguished three dimensions of AI support in sciences: providing hidden information, creative support, and support of scientific understanding. The latter would require the AI system to convey the reasoning process to the scientist by using adequate logical relations and visualisation of the underpinnings of the reasoning process. AI systems would need to recognise the implications of a model or theory if applied to another system and be able to convey the derived consequences to the human researcher. A first step has been taken by using causal inference in machine learning to develop explainable AI (Kuang et al., 2020). However, such an "agent of understanding" has not yet been described in the literature.

Philosopher of natural sciences Henk de Regt (2022) raise the question of whether understanding is based just on factual knowledge. Apart from the changing nature of knowledge due to paradigm shifts and new revolutionary insights, he emphasised that – besides the propositional (and codifiable) knowledge – implicit skills play a key role in understanding. Creativity relies on inspiration to apply existing knowledge to a new domain. However, there are other aspects to consider.

First, scientific reasoning relies on assumptions. The use of AI presupposes a mechanistic mindset that assumes that the human brain functions according to recognisable rules and a sophisticated but understandable mode of action, or, as Yuval Harari assumed, "millions of neurons calculating probabilities within a split second" (Harari, 2018, p. 20).

Second, generating reasonable explanations for observed or measured phenomena requires familiarity with the context of the explained phenomenon and a distinct sensitivity to anomalous behaviour of another. Whether AI can detect and evaluate context in a meaningful way to identify anomalies depends on the algorithmic implementation. Apart from fuzzy logic, computational algorithms rely on discrete boundaries that limit their interpretive power.

Third, finding relationships between phenomena involves conceptual reduction: a phenomenon is reduced to its explanatory aspect in its given context. Such simplification is the basis for building models and theories, which are essentially nothing more than conceptually reduced abstractions. Generalisation and abstraction, both involving simplification, require AI algorithms to identify features of the data that are characteristic to the research problem. This is hard to imagine without constant interaction between the researcher and the AI system.

Fourth, the reduction inherent to explanatory models necessarily limits the degree of freedom of explanation. American philosopher Michael Friedman (1974) pointed out the difference between explanation and understanding. In his view, merely relating new phenomena to known ones limits our ability to gain a comprehensive picture. He emphasises the need for abstraction, but also points out that it is of limited value for a comprehensive understanding. Therefore, deductive-nomological approaches limit understanding because the result of an experiment or observation is expected. Perhaps it is just a matter of asking nature the right questions. An electron microscope cannot provide an image of an electron because of its design. When scientific investigations do not produce the desired result, it is either because of inadequate assumptions or because scientists are asking nature the wrong questions by limiting the experimental or theoretical conditions according to an oversimplified model.

Finally, the explainability of theories or models is a matter of complexity of their underlying systems. While AI can identify information not apparent to scientists in complex data, can generalise observations, and can reach a certain level of abstraction, it seems only complement human intelligence.

The Sky is Not the Limit

If the machine is to participate in the research process, it is expected that – as with any human scientist – its thought process is reproducible and the underlying data valid. AI would then be just another computational tool: it would accelerate scientific endeavours but only marginally change their conceptual foundations. In fact, AI might be barely more than machine learning that makes science faster but not better. Jordan (2019, p. 3) simply describes it as a "new branch of engineering". Although supported by several authors (e.g. Candès et al., 2019) and criticised by others (e.g. Grosz, 2019), Jordan raises an important question: Can AI really reach or even surpass the expectations of the scientific community?

AI is (still) a simplified imitation of human intelligence. Even when scientists develop intelligent systems, they create them using the human thought process as a blueprint. In other words, the sky is not the limit, humans are. One of the most widely accepted definitions of creativity comes from cognitive scientist Margaret Boden. She argues that creativity is an ability characterised by an element of surprise and involves, among other things, awareness, emotional response, and intentionality (Boden, 2004). Although controversial (e.g. Botha, 2023), these abilities can hardly be attributed to current AI technologies, at least in the contemporary understanding of AI. There are good reasons to be sceptical about the capabilities of AI, as Marcus and Davis (2019) have pointed out. The hype on artificial intelligence is driven by an industry that expects exorbitant revenues from it and is propelled by uncritical media and its most enthusiastic supporters who envision a bright future. Currently, AI systems create axiomatic proofs, deductive reasoning, inductively generate hypotheses that they then test statistically, find hidden information, and perform all of this with massive amounts of data and unmatched speed. But should that really be all?

Surpassing Cognitive Boundaries

AI not just supports scientists in their creative tasks and improves the quality and outcomes of research. It can even lead to gaining previously unattainable insights and advance human critical thinking and capabilities of abstraction in unprecedented ways. But what if there is more to it? What if AI brings forth a new kind of creativity that surpasses cognitive boundaries? What if AI gains a level of abstraction that is unattainable to humans? What are the limits then?

In the worst case, it ends with scientists taking the credibility of results generated by AI for granted without questioning the underlying reasoning mechanisms anymore. Such black-box mentality is common in everyday life when humans carelessly use computers or smartphones without really understanding their technical underpinnings or imagining the implications of their digital actions for their privacy and security. The same may apply to scientists and even AI engineers may be overwhelmed by the complexity of an AI-based thought process. In any case, close collaboration between scientists and AI engineers in the development of AI-based science would be highly advisable. Explainability is a basic requirement if AI should play its part as creative scientist. However, as AI reaches a level of creativity and abstraction incomprehensible to humans, the concept of explainability may become absurd.

Another scenario would be if AI's creativity reaches a level that converges to human creativity, including – in Boden's sense – self-awareness, emotional response, and intentionality. Although there are doubts about whether an artificial general intelligence (AGI) can be realised (see, e.g. Fjelland, 2020), there are promising approaches that at least hold out the possibility of generating such systems. One example is AI-generating algorithms that use meta-learning architectures to autonomously learn the building blocks of AGI systems (Clune, 2019). Another is a third-generation AI using a triple-space model to bridge the gap between a continuous feature space, which describes the characteristics of observed or measured data, and a discrete semantic space that provides the meanings of data and contributes to explainability (Zhang, 2023). Pei et al. (2019) even proposed a chip supporting machine-learning algorithms as well as neuroscience-based models simultaneously to achieve AGI.

The mere fact that such ideas are being worked on raises some concern: it may just be a matter of conservatism, lack of adaptability, and a fair amount of discomfort that keeps scientists from considering the possibility of generating a creative, self-aware, intentional, or even sentient AI.

If successful, such an approach would lead to a paradigm shift that goes well beyond Kuhn's (1970) understanding of the term. Unlike quantum theory, which changed the perception of physical reality, or the liberalisation in the social sciences that enhanced the view on society, the emergence of an inherently creative AI would go beyond a mere "methodological sophistication" (Anand et al., 2020, p. 1655). Not only would it challenge the epistemology and the self-conception of researchers, but it could have unprecedented implications for society.

A Call for Early Action

Science is about understanding the world and finding the rules behind what we can perceive or imagine. However, scientists exist not in isolation within an ivory tower. Their discoveries significantly influence society, shaping our self-perception, values, and norms. While AI can open unimagined possibilities for developing our world view, there is a fine line we could cross with artificial creativity to end with AGI. That would call into question the role of the scientist as researcher, the role of society as knowledge bearer, and the role of humans as the measure of intelligence.

The ethical discourse of the consequences of advanced technology (for extensive reviews, see Stahl, 2021; McLean et al., 2021) often misses an important aspect that requires a more imaginative perspective. If AI could further reach a level of "artificial" consciousness, intentionality, or even sentience, we need to seriously reconsider our perception of AI systems as "artificial". We might engage in creating a new form of intelligence, a post-human species. In this case, we must judge ourselves by how we deal with this species, what rights we grant it, and what responsibility we take on us by challenging the established anthropocentric world view (see also the discussion of Trujillo, 2023).

Humanity usually thinks too late about the consequences of technological developments. We discover nuclear fission before we can prepare for its dramatic consequences. We learn to tame it before we experience the worst-case scenario of a meltdown. Similar lack of foresight in contemplating the implications of creative AI could lead to societal disruptions surpassing any imaginable conflict scenario. Without further exploring the implications of incorporating creative AI into epistemology, neither scientists nor society will have the slightest idea of whether they will thrive or perish in this transition. It is our responsibility to proactively address the consequences of such a paradigm shift – and this time before we are overtaken by their repercussions.

Future Research Agenda

The scenarios outlined above, regardless of their feasibility, underscore the potential challenges that arise when AI surpasses human capabilities in the epistemological process and their consequences in shaping our understanding of the sciences. Consequently, they demand the full attention of the research community. Delving into the following questions would serve a first exploratory analysis:

RQ1: Do AI systems, with their capacity for creativity and abstraction in the research process, mirror human abilities, or do they surpass them?

It is necessary to identify characteristics of "creativity" and "abstraction" of AI systems and their differences from human abilities. This requires a review of achievements of AI systems exhibiting creativity and abstraction in research scenarios. Depending on the differences, a definition is required that goes beyond current approaches to formalising creativity (e.g. Simonton, 2021) and abstraction (e.g. Yee, 2019).

RQ2: What are the epistemological consequences when creativity and abstraction of AI systems surpass human abilities?

Models and theories are conceptually reduced abstractions and play a fundamental role in sciences. Creative capabilities of AI may lead to systems that surpasses human's level of abstraction, thereby creating models that are hard to evaluate. Research practitioners and AI engineers need to investigate minimal conditions, general capabilities, and limitations of explainability of such systems. Furthermore, the questions of how knowledge is then generated, validated, communicated, and made available must be discussed in a wider science community.

RQ3: What contemporary concepts of self-generating AI leading to general AI are being employed in research?

A comprehensive systematic review of self-generating AI systems in research and development is needed, followed by a critical assessment of their feasibility, their capabilities, and their contribution to the complexity of explanation and understanding.

RQ4: What are ethical consequences of AGI and its use in the epistemological process?

If there is a possibility that AI creativity will approach Boden's definition of human creativity, then ethical considerations must extend beyond the purely technological implications of AGI to the creation of a new form of intelligence and its implications for science and society.

References

Abiodun, O. I., Omolara, A. E., Dada, K. V., Umar, A. M., Linus, O. U., Arshad, H., Kazaure, A. A., Gana, U., & Kiru, M. U. (2019). Comprehensive review of artificial neural network applications to pattern recognition. *IEEE Access*, 7, 158820–158846. https://doi.org/10.1109/ACCESS.2019.2945545

Anand, G., Larson, E. C., & Mahoney, J. T. (2020). Thomas Kuhn on paradigms. *Production and Operations Management*, *29*(7), 1650–1657.

Boden, M. A. (2004). *The creative mind: Myths and mechanisms*. Routledge.

Botha, C. F. (2023). Creativity and AI: A response to Boden. 2022, 366, 204. In R. Hakli, P. Mäkelä, & J. Seibt (Eds.), *Social robots in social institutions: Proceedings of robophilosophy 2022* (pp. 204–210). IOS Press.

Briganti, G. (2022). On the use of Bayesian artificial intelligence for hypothesis generation in psychiatry. *Psychiatria Danubina*, *34*(8), 201–206.

Candès, E., Duchi, J., & Sabatti, C. (2019, July 1). Statistics and the oncoming AI revolution. *Harvard Data Science Review*, *1*(1). https://doi.org/10.1162/99608f92.ff7ca64e

Clune, J. (2019). AI-GAs: *AI-generating algorithms, an alternate paradigm for producing general artificial intelligence. arXiv*, arXiv:1905.10985v2. https://arxiv.org/pdf/1905.10985.pdf

de Regt, H. W. (2022). Can scientific understanding be reduced to knowledge? In I. Lawler, K. Khalifa, & E. Shech (Eds.), *Scientific understanding and representation: Modeling in the physical sciences* (pp. 17–32). Routledge.
El Naqa, I., Haider, M. A., Giger, M. L., & Ten Haken, R. K. (2020). Artificial intelligence: Reshaping the practice of radiological sciences in the 21st century. *The British Journal of Radiology*, *93*(1106), 20190855. https://doi.org/10.1259/bjr.20190855
Figoli, F. A., Mattioli, F., & Rampino, L. (2022). *Artificial intelligence in the design process: The impact on creativity and team collaboration*. FrancoAngeli S.R.L.
Fjelland, R. (2020). Why general artificial intelligence will not be realized. *Humanities and Social Sciences Communications*, *7*(1), 1–9. https://doi.org/10.1057/s41599-020-0494-4
Friedman, M. (1974). Explanation and scientific understanding. *The Journal of Philosophy*, *71*(1), 5–19.
Goodfellow, I. (2017, April 3). *NIPS 2016 tutorial: Generative adversarial networks*. Cornell University. https://doi.org/10.48550/arXiv.1701.00160
Griffith, T. L., & Mangla, U. (2023). Virtual collaboration: Human foundations augmented by intelligent technology. In L. L. Gilson, T. O'Neill, & M. T. Maynard (Eds.), *Handbook of virtual work* (pp. 41–66). Edward Elgar Publishing.
Grimmer, J., Roberts, M. E., & Stewart, B. M. (2021). Machine learning for social science: An agnostic approach. *Annual Review of Political Science*, *24*, 395–419.
Grosz, B. J. (2019, July 2). The AI revolution needs expertise in people, publics, and societies. *Harvard Data Science Review*, *1*(1). https://doi.org/10.1162/99608f92.97b95546
Harari, Y. N. (2018). *21 lessons for the 21st century*. Random House Publishing Group.
Huang, J., & Tan, M. (2023). The role of ChatGPT in scientific communication: Writing better scientific review articles. *American Journal of Cancer Research*, *13*(4), 1148–1154.
Jordan, M. I. (2019, December 11). Artificial intelligence – the revolution hasn't happened yet. *Harvard Data Science Review*, *1*(1), 1–9. https://doi.org/10.1162/99608f92.f06c6e61
Kara, H. (2020). *Creative research methods – a practical guide*. Policy Press and Bristol University Press.
King, R., Peter, O., & Courtney, P. (2023). Artificial intelligence in science: Robot scientists: From Adam to Eve to genesis. In *Artificial intelligence in science: Challenges, opportunities and the future of research* (pp. 129–139). OECD Publishing. https://doi.org/10.1787/d89b39b6-en
Krenn, M., Pollice, R., Guo, S., Aldeghi, M., Cervera-Lierta, A., Friederich, P., dos Passos Gomes, G., Häse, F., Jinich, A., Nigam, A., Yao, Z., & Aspuru-Guzik, A. (2022). On scientific understanding with artificial intelligence. *Nature Reviews Physics*, *4*(12), 761–769.
Kuang, K., Li, L., Geng, Z., Xu, L., Zhang, K., Liao, B., Huang, H., Ding, P., Miao, W., & Jiang, Z. (2020). Causal inference. *Engineering*, *6*(3), 253–263. https://doi.org/10.1016/j.eng.2019.08.016
Kuhn, T. (1970). *The structure of scientific revolutions* (2nd ed.). The University of Chicago Press.
Lauro, C., Pereira, R., Brandão, L., & Davim, J. (2015). Design of experiments – statistical and artificial intelligence analysis for the improvement of machining processes: A review. In J. P. Davim (Ed.), *Design of experiments in production engineering* (pp. 89–107). Springer International.
Ludwig, J., & Mullainathan, S. (2023). *Machine learning as a tool for hypothesis generation*. National Bureau of Economic Research. https://doi.org/10.3386/w31017
Marcus, G., & Davis, E. (2019). *Rebooting AI: Building artificial intelligence we can trust*. Pantheon Books.

Mazzone, M., & Elgammal, A. (2019). Art, creativity, and the potential of artificial intelligence. *Arts*, *8*(1), 26–35. https://doi.org/10.3390/arts8010026
McCormack, J., Hutchings, P., Gifford, T., Yee-King, M., Llano, M. T., & d'Inverno, M. (2020). Design considerations for real-time collaboration with creative artificial intelligence. *Organised Sound*, *25*(1), 41–52.
McLean, S., Read, G. J., Thompson, J., Baber, C., Stanton, N. A., & Salmon, P. M. (2021). The risks associated with artificial general intelligence: A systematic review. *Journal of Experimental & Theoretical Artificial Intelligence*, 1–15.
Oquendo, M. A., Baca-Garcia, E., Artes-Rodriguez, A., Perez-Cruz, F., Galfalvy, H. C., Blasco-Fontecilla, H., Madigan, D., & Duan, N. (2012). Machine learning and data mining: Strategies for hypothesis generation. *Molecular Psychiatry*, *17*(10), 956–959.
Pei, J., Deng, L., Song, S., Zhao, M., Zhang, Y., Wu, S., Wang, G., Zou, Z., Wu, Z., He, W., Chen, F., & Shi, L. (2019). Towards artificial general intelligence with hybrid Tianjic chip architecture. *Nature*, *572*(7767), 106–111.
Ray, P. P. (2023). ChatGPT: A comprehensive review on background, applications, key challenges, bias, ethics, limitations and future scope. *Internet of Things and Cyber-Physical Systems*, *3*, 121–154. https://doi.org/10.1016/j.iotcps.2023.04.003
Salvagno, M., Taccone, F. S., & Gerli, A. G. (2023). Can artificial intelligence help for scientific writing? *Critical Care*, *27*(1), 1–5.
Shao, Z., Yuan, S., & Wang, Y. (2020). Institutional collaboration and competition in artificial intelligence. *IEEE Access*, *8*, 69734–69741. https://doi.org/10.1109/access.2020.2986383
Simonton, D. K. (2021). Scientific creativity: Discovery and invention as combinatorial. *Frontiers in Psychology*, *12*, 721104. https://doi.org/10.3389/fpsyg.2021.721104
Stahl, B. C. (2021). *Artificial intelligence for a better future: An ecosystem perspective on the ethics of AI and emerging digital technologies*. Springer Nature.
Topalovic, M., Das, N., Burgel, P. R., Daenen, M., Derom, E., Haenebalcke, C., Janssens, W., Kerstjens, H. A., Liistro, G., Louis, R., & Ninane, V. (2019). Artificial intelligence outperforms pulmonologists in the interpretation of pulmonary function tests. *European Respiratory Journal*, *53*(4).
Trujillo, J. (2023). A phenomenological reply to Gary Marcus and Ernest Davis's rebooting AI: Building artificial intelligence we can trust: Contributing to the creation of general artificial intelligence. *Creativity Studies*, *16*(2), 448–464.
Van Dis, E. A., Bollen, J., Zuidema, W., van Rooij, R., & Bockting, C. L. (2023). ChatGPT: Five priorities for research. *Nature*, *614*(7947), 224–226.
Wang, H., Fu, T., Du, Y., Gao, W., Huang, K., Liu, Z., Chandak, P., Liu, S., Van Katwyk, P., Deac, A., Anandkumar, A., & Zitnik, M. (2023). Scientific discovery in the age of artificial intelligence. *Nature*, *620*(7972), 47–60.
Wang, S., Scells, H., Koopman, B., & Zuccon, G. (2023). Can ChatGPT write a good boolean query for systematic review literature search? In *Proceedings of the 46th international ACM SIGIR conference on research and development in information retrieval* (pp. 1426–1436). The Association for Computing Machinery.
Xames, M. D., & Shefa, J. (2023). ChatGPT for research and publication: Opportunities and challenges. *Journal of Applied Learning & Teaching*, *6*(1), 390–395. https://doi.org/10.37074/jalt.2023.6.1.20
Yee, E. (2019). Abstraction and concepts: When, how, where, what and why? *Language, Cognition and Neuroscience*, *34*(10), 1257–1265.
Yuan, S., Shao, Z., Liang, Y., Tang, J., Hall, W., Liu, G., & Zhang, Y. (2020). International scientific collaboration in artificial intelligence an analysis based on web

data. In *Proceedings of the 12th ACM conference on web science* (pp. 69–75). Association for Computing Machinery.

Zhai, X., Chu, X., Chai, C. S., Jong, M. S., Spector, M., Liu, J., Yuan, J., & Li, Y. (2021). A review of artificial intelligence (AI) in education from 2010 to 2020. *Complexity*, *2021*, 1–18. https://doi.org/10.1155/2021/8812542

Zhang, B. Z. (2023). Toward the third generation artificial intelligence. *Science China Information Sciences*, *66*(2), 121101. https://doi.org/10.1007/s11432-021-3449-x

Chapter 16

Recent Advancements and Use Cases on Co-Creation, Procedural Content Generation, and Computational Creativity in Video Games

Elif Surer

Introduction

Human creativity has always been an essential part of human existence, and as Colton and Wiggins (2012, p. 21) emphasize in their considerable work:

> *As a society, we are jealous of our creativity: creative people and their contributions to cultural progression are highly valued. Moreover, creative behaviour in people draws on a full set of intelligent abilities, so simulating such behaviour represents a serious technical challenge for Artificial Intelligence research.*

There have been numerous definitions of creativity, which is a very complex, diversified, and abstract concept that is difficult to quantify, solidify, and evaluate. Creativity is defined by Margaret Boden (2010, p. 29) as "*the ability to come up with ideas or artifacts that are new, surprising, and valuable*".

Given how challenging it is to identify and study human creativity, "computational creativity" is another difficult topic worth decoding. In this chapter, we will utilize Colton and Wiggins's definition as a basis for the examples, use cases, and terminology:

> *The philosophy, science and engineering of computational systems which, by taking on particular responsibilities, exhibit behaviours that unbiased observers would deem to be creative.*
>
> (Colton & Wiggins, 2012, p. 1)

While creativity is commonly defined under the "lonely genius" form – where the creative task is thought to be performed in isolation only – collaboration, brainstorming, and co-creation have been the catalysts in creativity, whether it is human or computational.

The goal of this chapter is to present a survey on the co-creation phenomenon in video games. While doing that we will briefly discuss the recent advancements in artificial and computational intelligence in games,

DOI: 10.4324/9781003453901-20

procedural content generation (PCG), and computational creativity. In the literature, all of these concepts and definitions overlap in some ways, and some of these phrases or definitions are not mutually exclusive: Computational creativity, for example, can be classified under artificial and computational intelligence whereas PCG can also be classified as computational creativity. In their study, Colton and Wiggins define computational creativity as a subfield of artificial intelligence (AI) (Colton & Wiggins, 2012). Our goal here is not to propose a new taxonomy or start a debate about how to categorize these concepts but to cover the definitions, ideas, and literature as broadly as possible, highlight each of their distinct features, and familiarize the reader with all these terms and recent studies performed within these domains.

In the following sections, we will first provide the main definitions, preliminary information, and background to serve as a foundation for the subsections that follow, and then we will provide exemplary studies and use cases of artificial and computational intelligence in games, with examples contributing to the advancements in co-creation, PCG, and computational creativity. Finally, in the "Future Research Agenda" section, we will look for answers in the recent literature to three important questions: (1) What are the key problems of video game co-creation during the development phase? (2) What are the primary problems of using artificial and computational intelligence in game development? and (3) What are the main debates on the ethics of AI in game development? These are thoroughly discussed on the ethics of AI and intellectual property rights and whether a computer can be creative.

Preliminaries and Background

Video games are digital games that include, among other things, game mechanics, a story, a user interface, audio, and levels and are primarily designed for entertainment purposes. Serious games are a subset of video games that have additional purposes other than entertainment only, such as education, rehabilitation, and training. The art, engineering, and craft (Schell, 2008) of creating gameplay with a wide range of tasks such as game mechanics, game scenarios, game levels, user interface, audio, and AI is known as game development. Given that creating game content is a hard, time-consuming, and expensive process, most game studios employ procedural content creation to generate high-quality content with parameterization and automation.

In general, hundreds of individuals with expertise in character design, animation, audio, narration, and level design work on AAA (Triple-A) games that are high-profile and high-budget since game design is a complex endeavour that calls for a wide range of skills. Human expertise is incredibly expensive; as Shaker et al. (2016) emphasize in their book, humans are not particularly effective in describing complex tasks that can be carried out so quickly in an algorithmic method. For instance, Alpaydin (2020) notes that although humans are very skilled at recognizing the voices and/or faces of their loved ones, it is difficult to explain this phenomenon to a computer using an algorithm. Since humans have spent countless hours becoming "trained" to recognize faces and voices, it is simple for us to do so even in less-than-ideal circumstances. Therefore, in terms of efficiency and effectiveness, automated technologies that can mimic and/or improve human expertise are advantageous.

Creativity and Computational Creativity

The improvement of human creativity is facilitated using artificial and computational intelligence in games as a collaborator, an enhancer, and a "muse". The application of artificial and computational intelligence enables people to understand the "why" behind a variety of agent-based techniques, enabling them to employ those methods when playing and testing video games (Shaker et al., 2016).

Game development, player behaviour analysis, usability analysis, game testing, and game AI are just a few of the topics covered in video game research. Playtesting, often known as game testing, involves playing a game and checking thoroughly for bugs. For many years, game AI has been a significant research subject, serving as a testbed for advanced and complicated AI algorithms. Game AI has long been an important research domain in which agents that can play or test video games and create content have served as an important testbed not only for game research but also for advanced AI algorithms covering various scientific domains such as reinforcement learning (RL), robotics, and biology (Risi & Preuss, 2020). Player modelling seeks to interpret the player's behaviour to provide a personalized design and adaptive content based on the preferences of that specific user.

Most artificial and computational intelligence research focuses on player modelling, game analytics, content development, and learning. Several machine learning techniques are used in game analytics to give a detailed examination of game data such as retention, playtime duration, unique game dynamics, and player preferences. Content generation will be covered in depth under PCG, where the objective is to create automated and parametric game content. Learning is mostly accomplished through machine learning, and a subcategory of machine learning – RL – is also widely employed in video game research. Game development requires a wide range of skill sets, including the creation of images, sound, levels, and stories. Liapis (2021) presents six areas of creativity in video games – graphics, rules, gameplay, audio, and story – and addresses the application of AI in the collaboration of professionals from different backgrounds, like designers and artists, creating specialized content.

Machine learning is defined as "*programming computers to optimize a performance criterion using example data or past experience*" by Alpaydin

(2020, p. 3). While machine learning covers association, supervised learning, unsupervised learning, and RL, deep learning is another subcategory of AI composed of "deep" layers of a network that focus on the task of extracting high-level features from large amounts of data. In RL, an agent tries to learn the surrounding environment to choose the actions that will maximize its cumulative reward during a sequential task. The most common applications of RL are in video games, bioinformatics, and robotics. RL is also classified according to whether the environment's model is known, whether the environment is fully or partially observable, for example. One of the key challenges of RL is fine-tuning the reward function to encourage or discourage the agent's decision.

The task of generating content using AI is known as generative AI. It has grown in importance as a result of advances in deep learning, particularly generative adversarial networks, and the recent success of large language models. Lv (2023) emphasizes the relevance of generative AI in content generation innovation in the Metaverse age, citing the impact of recent breakthroughs in large language models, such as ChatGPT. Lv argues that ChatGPT has the ability to transform the way information is generated and presented, as well as provide suggestions and insights into how to employ generative AI more efficiently while producing creative content.

In game research, generative AI is commonly utilized to generate games, game levels, content, and audio.

Artificial and computational intelligence has mostly been used in game creation, level design, game testing, and content generation. The line between computational intelligence and computational creativity has become more blurred in the literature (Ventura, 2016). Ventura (2016) proposes that computational creativity will be the next breakthrough in game design and provides a full definition of the concept as well as a discussion of how computational creativity might be implemented into the game design process. To do this, the author proposes new incentives to encourage collaboration between computational intelligence and game designers. Under the "artificial and computational intelligence" term, various aspects of the application of machine learning techniques, procedural content creation, and generative AI in video games are covered.

Co-Creation and Computational Creativity

The task of creating together is known as co-creation. Co-creation between a human and a machine has previously been employed in a variety of tasks, including game development, content generation, and game testing. Before we get into the "how" of co-creation, it is worth noting that the "when" of co-creation determines the sharing or division of tasks. According to Yannakakis et al. (2014) in their substantial "mixed-initiative" study, the timing of co-creation, such as before the creation process (setting the parameters), during the creation process (both humans and computers participating proactively), and after the creation (editing a procedurally generated 3D assets), also determines the ratio of the co-creation, and whether it is simultaneous, sequential, or in different turns. In their study, Yannakakis et al. (2014) offer the following questions to begin an important dialogue on co-creation and creativity:

> *Do we co-create when we design and create with a machine, or can a machine enhance our creativity as human creators? When does this kind of co-creation encourage the co-creation of both robots and humans?*
> (p. 1)

Their findings reveal that their proposed framework, Sentient Sketchbook, boosts human creativity and actualizes mixed-initiative co-creation.

Several studies on co-creation have been conducted, and generative AI is the dominating topic in co-creation. In a recent study, Yetkin et al. (2023) transfer RL agents and gameplay concepts to develop unique outcomes in the architectural domain. A unique technique based on RL is explained, allowing designers to leverage intelligent agents to their advantage when investigating the potential through data collecting and data processing. The difficulty of collaboration is presented using both top-down and bottom-up approaches. The top-down technique involves selecting a scenario in which the designer seeks to handle a house design issue, beginning with verifying that the total area and distribution of housing unit types meet the general requirements. The bottom-up technique selects a case in which the designer intends for the same problem by meeting the specific needs of room space and relationships. Both scenarios are based on an RL technique that allows the user to interact with the RL algorithm.

Cook and Colton (2018) investigate the concept of ongoing creativity by demonstrating how ANGELINA, a game engine that allows players to build new games through prompts, was modified to be an "always-on" system. The authors review ANGELINA's history before comparing their new approach to previous iterations. The presence of a computationally creative system is defined as the impact it has on its surroundings and vice versa. The authors describe how continuing innovation can improve a system's presence by allowing it greater creative flexibility, framing possibilities, and opportunities to grow.

The parametric and automated production of content, particularly in video games, is known as PCG. PCG is mostly utilized in the creation of landscapes, levels, and audio. Because generating diverse and automated content by human designers takes a long time, and human knowledge is time-consuming and expensive, PCG is a tremendous help for game designers. Using PCG allows small- and mid-sized game companies to compete and thrive in the competitive game design market by allowing them to develop diversified content in less time, with fewer people, and with fewer resources.

PCG was heavily used in the 2016 game *No Man's Sky*. PCG has also been a significant research area, with multiple books and scientific articles published on the subject, as well as numerous competitions held at game conferences such as IEEE Conference on Games and Association for the Advancement of Artificial Intelligence Conference. PCG has lately been utilized in conjunction with machine learning and RL. PCG with machine learning (PCGML) has primarily been utilized for game content creation purposes (Summerville et al., 2018), whereas PCG with reinforcement learning has primarily been used for level generation (Khalifa et al., 2020). Summerville et al. (2018) investigate PCGML, which is the process of creating game content training machine learning models on previously released material. The authors discuss the use of machine learning (rather than previous methods such as search-based, solver-based, or constructive methods) while focusing on practical game content, such as levels, game maps, storytelling, and cards in collectible card games. Khalifa et al. (2020) investigate the use of RL to train level-designing agents, which presents a revolutionary approach to content creation in video games by framing level building as a game and learning the content generator. By viewing the design challenge as a sequential task, the authors apply RL to learn the next action to optimize the anticipated final level quality.

Exemplary Studies and Use Cases of Co-Creation and Computational Creativity

Computer-assisted creativity and human-centred AI systems are powerful forces behind computational creativity; according to the authors of the book on AI, arts, and computational creativity (Machado et al., 2021), creative and artistic behaviour is one of the main challenges facing AI research. The book examines computational creativity, more particularly the creation of systems that can foster human creativity and artistic ability or that act in an artistic manner. In areas including gaming, music, 3D environments, and the visual and performing arts, the authors examine and consider current trends, identify significant barriers and opportunities, and propose novel solutions.

Guzdial et al. (2018) use machine learning methods to enable co-creation during level design. In their procedural level generation through machine learning (PLGML) study, they present a system for collaborative-level design with a PLGML agent. To analyse this concept, they provide data from user studies and conclusions from a comparison of PLGML methodologies.

The term "co-creation" was later utilized by Dharna et al. (2020), who also create game levels and game-playing agents. A co-evolutionary system called the paired open-ended trailblazer (POET) algorithm simultaneously creates new environments and agents solving and navigating these environments. This study proposes the POET-inspired Neuroevolutionary System for KreativitY (PINSKY) in games, which jointly develops a variety of video game levels and agents that play these levels. This framework uses the general video game AI (GVGAI) architecture to support the co-creation of levels and agents for the games such as *Solar Fox* and *Zelda*, which are two-dimensional Atari-style games. As a result of their findings, an interesting new research topic harmonizing PCG and artificial life has been made possible.

A comparative analysis based on the data from human testers was used by Ariyurek et al. (2021) to assess the results of creating human-like game testers. The authors also used the above-mentioned GVGAI environment for testing while providing a novel methodology for automating video game testing that employs tester agents. The authors define two types of agents: synthetic and human-like, as well as two distinct ways of creating them. Their bug-finding agents are extracted from Monte Carlo Tree Search and RL agents. The synthetic agent employs test goals generated from game scenarios to investigate the impact of unexpected game interactions. After collecting 427 trajectory data from human testers, Ariyurek et al. (2021) built three different games each having 12 levels and 45 bugs using the GVGAI framework. According to their findings, synthetic and human-like bots can spot bugs just as well as human testers.

Uricchio and Cizek (2023) conducted an in-depth examination of co-creating with AI. They contend that co-creation offers a practical strategy for exploring expressive skills and potential forms of agency in systems that have already been identified as prospective candidates for consciousness. In this study, the authors investigate the prospect of collaboration between non-human systems, notably systems powered by AI. They interviewed more than 30 artists, writers, curators, and programmers to understand more about how they connect with the AI systems they employ in their work. For the benefit of both human and non-human futures, their responses frequently represent a wider range of co-creation, broadening the social discussion and challenging problems of agency and non-agency, technology, and control.

Cook et al. (2019) present a study and taxonomy of computational creativity in light of current advances in the field. Building computationally creative software necessitates understanding the concept of framing computationally created artefacts, which entails providing a narrative context for the software's actions and objectives. Their work presents the first assessment of framing in computational creativity, as well as a taxonomy of framing components such as motivation, implementation, and rendering. They also address future directions for framing and its importance in the future progress of the field.

Guzdial et al. (2019) showcase the numerous roles AI performs during game-level design and the impact it has on its human designers. The authors build a Super Mario Bros-like game-level creation tool with an integrated AI-level designer to examine how the designers practice, view, and integrate AI in the creative process. In this study, they present the design of an intelligent tool named Morai Maker, which is tested in two trials using mixed methods and a total of over 100 participants.

Future Research Agenda

While co-creation and computational creativity provide numerous solutions and spectacular outcomes, there are a number of issues, challenges, and ongoing debates on the use of artificial and computational intelligence. This subchapter intends to present questions that emphasize these issues related to the video game development while also encouraging fresh conversations, brainstorming, and moves towards solutions.

RQ1: What are the key problems of video game co-creation during the development phase?

One of the key features of co-creation is the feeling of control and autonomy. The human designers and/or developers' sense of control and autonomy may shift depending on the phase of the co-creation process. For example, as indicated in Yannakakis et al.'s (2014) study, a human designer would feel entirely in control while selecting the parameters for a PCG algorithm, given that the parameters are chosen by them, and that the outcome is quite predictable. Again, towards the end of an automated content generation cycle, a human designer might feel entirely in control of the generated content while editing the final outcome. The sensations of agency and control are not lost when the cooperation is turn-based or one-sided, but they are challenged when the co-creation process is reciprocal, proactive, and uncertain. However, having the sensation of losing control may be preferred on some occasions. Detaching from the problem at hand, sleeping on a solution, and brainstorming in a free manoeuver, as shown in Harrison's (2006) book *Ideaspotting*, can help improve creativity. Given that the frequency, depth, and length of co-creation may be crucial elements in human–agent collaboration, the granularity of the co-creation process can be an intriguing topic to investigate.

RQ2: What are the primary problems of using artificial and computational intelligence in game development?

One of the most difficult issues in employing artificial and computational intelligence in games is creating unique and novel content. For some researchers working in the creative domain, style transfer, the parametric and automated generation of outcomes, and level generation are still based on the initial human-created content or parameterized formula, so the generated outcomes are "derivations" of human creativity rather than truly creative. Some researchers claim that creativity is a human quality, and as long as a computer is behind the generated content, it cannot be called creative (Inie et al., 2023).

The "black-box" nature of machine learning algorithms makes it difficult to predict the expected outcome and reduces the designer's agency. The explainability of machine learning systems has recently become a hot topic in research, as knowing the machines' "line of thought" would greatly improve the design of the experimental setups. In Llano et al.'s study (2020), it is mentioned that human collaboration with systems is frequently limited to brief interactions in the computational creativity field, where systems' and humans' creative processes are performed independently, with little (or no) user involvement, and without analysing or discussing the decision-making procedures. A long and continuous idea exchange, comparisons to other works, minor tweaks, and alterations are all required for successful co-creation. Communication is an essential part of all these interactions.

Finally, AI-generated content necessitates additional monitoring in terms of playability, difficulty levels, and adaptability. The negative statements stressing the "machines' ruling of the world" and "AI taking our jobs" create resistance and hesitancy among the general public and reduce the diverse use of co-creation, particularly in creative industries. Ownership and intellectual property rights have been a source of debate on the collaboration with AI topic, and the same issues apply to collaboration with video game agents.

RQ3: What are the main debates on the ethics of AI in game development?

This question surely deserves at least another book chapter to be thoroughly examined, although the most recent study by Melhart et al. (2023) is a good place to start. There has been ongoing debate on the ethics of employing AI in games and the discussions have grown in importance as artificial and computational intelligence in games has advanced.

Inie et al. (2023) present a "participatory AI" strategy in which both humans and machines actively produce content. They also conduct interviews and a recent poll among creative persons and artists to learn about their viewpoints on generative AI in the creative sector. While some of the artists are concerned about the potential influence of AI, most of them see AI as a catalyst for their work, enabling greater outcomes and variety, particularly through co-creation. The "black box" nature of machine learning algorithms and their unpredictability, long periods of time for training data, extensive fine-tuning of reward functions, and requiring extra measures to evaluate the playability of the generated games and/or levels, originality, and applicability of the created content are the main challenges of using AI in games. There is still a long way to go in terms of legal and copyright issues, intellectual property, and ownership.

Finally, all these three questions have been presented here not to provide a definitive and final answer but to contribute to the discussion and familiarize the reader with the diversity and dynamism of the current ideas, challenges, and technologies without overlooking the complexity and beauty of the future that lies ahead of us.

References

Alpaydin, E. (2020). *Introduction to machine learning*. MIT Press.

Ariyurek, S., Betin-Can, A., & Surer, E. (2021). Automated video game testing using synthetic and humanlike agents. *IEEE Transactions on Games*, *13*(1), 50–67.

Boden, M. A. (2010). *Creativity and art: Three roads to surprise*. Oxford University Press.

Colton, S., & Wiggins, G. A. (2012, August). Computational creativity: The final frontier? In *Proceedings of the 20th European conference on artificial intelligence* (pp. 21–26). IOS Press.

Cook, M., & Colton, S. (2018). Redesigning computationally creative systems for continuous creation. In *International Conference on Computational Creativity 2018* (pp. 32–39). Association for Computational Creativity (ACC).

Cook, M., Colton, S., Pease, A., & Llano, M. T. (2019, June). *Framing in computational creativity-a survey and taxonomy* (pp. 156–163). ICCC.

Dharna, A., Togelius, J., & Soros, L. B. (2020, October). Co-generation of game levels and game-playing agents. *Proceedings of the AAAI Conference on Artificial Intelligence and Interactive Digital Entertainment*, *16*(1), 203–209.

Guzdial, M., Liao, N., Chen, J., Chen, S. Y., Shah, S., Shah, V., Reno, J., Smith, G., & Riedl, M. O. (2019, May). Friend, collaborator, student, manager: How design of an AI-driven game level editor affects creators. In *Proceedings of the 2019 CHI conference on human factors in computing systems* (pp. 1–13). CHI.

Guzdial, M., Liao, N., & Riedl, M. (2018). *Co-creative level design via machine learning. arXiv Preprint*. https://arxiv.org/abs/1809.09420

Harrison, S. (2006). *Ideaspotting: How to find your next great idea*. How Books.

Inie, N., Falk, J., & Tanimoto, S. (2023, April). Designing participatory AI: Creative professionals' worries and expectations about generative AI. In *Extended abstracts of the 2023 CHI conference on human factors in computing systems* (pp. 1–8). CHI.

Khalifa, A., Bontrager, P., Earle, S., & Togelius, J. (2020, October). PCGRL: Procedural content generation via reinforcement learning. *Proceedings of the AAAI Conference on Artificial Intelligence and Interactive Digital Entertainment*, *16*(1), 95–101.

Liapis, A. (2021). Artificial intelligence for designing games. *Artificial Intelligence and the Arts: Computational Creativity, Artistic Behavior, and Tools for Creatives*, *277*.

Llano, M. T., d'Inverno, M., Yee-King, M., McCormack, J., Ilsar, A., Pease, A., & Colton, S. (2020). Explainable computational creativity. In *International conference on computational creativity 2020* (pp. 334–341). Association for Computational Creativity (ACC).

Lv, Z. (2023). Generative artificial intelligence in the metaverse era. *Cognitive Robotics*, *3*.

Machado, P., Romero, J., & Greenfield, G. (Eds.). (2021). *Artificial intelligence and the arts: Computational creativity, artistic behavior, and tools for creatives* (pp. 3–381). Springer.

Melhart, D., Togelius, J., Mikkelsen, B., Holmgård, C., & Yannakakis, G. N. (2023). The ethics of AI in games. In *IEEE transactions on affective computing*. IEEE.

Risi, S., & Preuss, M. (2020). From chess and Atari to Starcraft and beyond: How game ai is driving the world of AI. *KI-Künstliche Intelligenz*, *34*, 7–17.

Schell, J. (2008). *The art of game design: A book of lenses*. CRC Press.

Shaker, N., & Nelson, M. J., & Togelius, J. (2016). Introduction. *Procedural Content Generation in Games*, 1–15.

Summerville, A., Snodgrass, S., Guzdial, M., Holmgård, C., Hoover, A. K., Isaksen, A., Nealen, A., & Togelius, J. (2018). Procedural content generation via machine learning (PCGML). *IEEE Transactions on Games*, *10*(3), 257–270.

Uricchio, W., & Cizek, K. (2023). Co-creating with AI. *Minnesota Review*, *2023*(100), 118–131.
Ventura, D. (2016, September). Beyond computational intelligence to computational creativity in games. In *2016 IEEE conference on computational intelligence and games (CIG)* (pp. 1–8). IEEE.
Yannakakis, G. N., Liapis, A., & Alexopoulos, C. (2014). Mixed-initiative co-creativity. In *Proceedings of the 9th conference on the foundations of digital games*. https://antoniosliapis.com/papers/mixed_initiative_co-creativity.pdf
Yetkin, O., Surer, E., & Gonenc Sorguc, A. (2023, October 18–20). *Collaborating with agents in architectural design and decision-making process: Top-down and bottom-up case studies using reinforcement learning*. EAI GOODTECHS 2023, 9th EAI International Conference on Smart Objects and Technologies for Social Good, Leiria, Portugal.

Chapter 17

Conclusions

"Embracing Human–AI Co-Creation: Navigating Through Opportunities and Challenges"

Francisco Tigre Moura

The chapters found in *Artificial Intelligence, Co-Creation and Creativity: The New Frontier for Innovation* provide relevant and timely contributions to the understanding of the transformative nature of artificial intelligence (AI). Overall, the book discussed the significant shifts in our understanding of creativity, innovation, and the role of human in creative tasks in a new era characterized by the impact of smart systems. The three main sections (*Part 1: Principles of AI and Creativity*; *Part 2: Critical Issues on Artificial Co-Creation*; *Part 3: Industry-Specific Discussions*) have, collectively, achieved the main aim of the book, which was to provide a holistic overview for readers from the perspective of scholars and practitioners. Furthermore, it has enabled important conclusions, which are summarized next.

Rethinking Creativity

Example: Chapters 1–3, 6, and 7.

A consensus emerging from the different chapters is the need to reconsider our established paradigm of creativity, the related concepts, theories, and frameworks. The traditional notion of humans as the main or sole creator in a creative process needs to change to a new perspective – one where the collaboration between humans and AI becomes not just desirable but essential. Creativity, in the age of AI, demands a redefinition that accommodates the symbiotic relationship between human intuition, feelings and motivation, with the impersonal and autonomous nature of algorithms.

Augmentation Unleashed

Example: Chapters 4–8, 11, and 14–16.

Another widely accepted view is the sense that AI should be seen not only as a tool, but as a powerful co-creator, capable of enabling extraordinary creative opportunities for human augmentation. The combination and application of human imagination with the computational power of AI is redefining different industry sectors and revealing uncharted territories of creativity.

DOI: 10.4324/9781003453901-21

From visual arts to music composition, from scientific research to gaming, AI acts as a catalyst, pushing the boundaries of what was once thought possible for humans to achieve alone or collectively in a short period of time.

The Evolving Role of Humans in Artificial Co-Creation

Example: Chapters 3–8, 10–12, and 14–16.

As we embrace the frontier of artificial co-creation, the intrinsic essence of human creativity undergoes a metamorphosis. Co-creating with AI is not a simple process; it transforms our perception of humans as creatives, deeply impacting our self-identity and sense of purpose. The traditional dichotomy of creator and creation blurs, as humans become "simply" partners in symbiotic collaborative processes with algorithms. This shift raises profound questions about authorship, ownership, psychological ownership, originality, and the nature of artistic expression in a landscape where the lines between human and machine contributions become increasingly indistinct. Currently, we still have many more questions than answers to these issues, and extensive further research is still required to address them.

Context Matters

Example: Chapters 7, 8, and 12–16.

The book has also shed light on the fact that the nature and implications of AI in co-creation are context-dependent. The rules and premises of one context do not necessarily apply to another. Whether it is in art, gaming, science, or education, the modes of co-creation vary significantly and so does the expectations of creators and receivers of the creative outputs. This emphasizes the need for continuous longitudinal interdisciplinary research that focuses on AI's role in creativity across diverse domains, enabling their permanent comparison. This will surely enable the maturing of understanding and application of artificial co-creation, acknowledging the specificities of each field.

Expecting the Unexpected in Artificial Co-Creation

Example: Chapters 1, 2, 6, and 7.

The only constant in the field of AI is a permanent and fast state of change. With the continuous evolution and relentless innovation in AI technology, humans are witnessing and constantly learning novel ways of co-creating with it. As algorithms become more sophisticated and adaptable, we must expect innovative and unexpected forms of artificial co-creation. Thus, the landscape of creative processes demands dynamic responses, permanent adaptation, and proactive behaviour by humans to deal with the potential of AI as a co-creative force.

Risks and Challenges

Example: Chapters 3, 10, and 15.

While the book has mostly focused on positive facets of AI in creativity, it is imperative to acknowledge the risks posed by this transformation. The potential risks and challenges posed by AI in creative processes are not to be taken lightly. Ethical concerns, bias in algorithms, loss of human agency, and the commodification of creativity are only a few examples of serious issues resultant from the collaboration between humans and AI. Given that this book has only addressed some of these complexities, future explorations must explore further the complexities of the AI-creative landscape.

Final Remarks

In conclusion, *Artificial Intelligence, Co-creation and Creativity: The New Frontier for Innovation* will hopefully provide a meaningful contribution to the current discussions in the field. Furthermore, all chapters have suggested timely future research questions, which can be addressed by scholars and students of different academic levels and also practitioners. Thus, we hope that it will serve as a relevant source of information and inspiration for future developments in the field.

Index

Note: Page numbers in *italic* indicate a figure and page numbers in **bold** indicate a table on the corresponding page.